Marketing of
Vegetables in India

Marketing of Vegetables in India

Vigneshwara Varmudy

2016
Daya Publishing House®
A Division of
Astral International Pvt. Ltd.
New Delhi - 110 002

© AUTHOR
First Impression, 2001
Reprinted, 2016

ISBN: 978-93-5130-657-3 (International Edition)

Published by : **Daya Publishing House®**
A Division of
Astral International Pvt. Ltd.
– ISO 9001:2008 Certified Company –
4760-61/23, Ansari Road, Darya Ganj
New Delhi-110 002
Ph. 011-43549197, 23278134
E-mail: info@astralint.com
Website: www.astralint.com

Laser Typesetting : **Classic Computer Services**
Delhi - 110 035

Digitally Printed at : **Replika Press Pvt. Ltd.**

PREFACE

The performance of agricultural sector in India since independence has been exemplary in terms of increased production, accounting for much higher growth in terms of production than the growth in population. However, availability of food alone is insufficient to overcome the basic problem of under-nourishment and hunger. In this regard, diversification of agriculture is an essential requirement. As vegetables are important components of diversification and contribute largely to solve the food and nutritional problems of the country, efforts are needed to increase the production, productivity and to improve the marketing conditions.

The vegetables produced in our country are much less than our requirement and serves per capita intake of only 135g, as against the requirement of 285 g, per capita per day for balanced diet. This is due to the prevailing pre- and post-harvest problems. These realities show that, India has to go a long way to accelerate the vegetable production, considering the demand for vegetables and to overcome the problem of socio-economic inequity. The country has failed to achieve the target of 100 million tonnes for the supply of 200 g. Vegetables per day per capita to our present population of one billion. The demand by 2030 will be around 250 million tonnes; to ahieve this herculean task, the vegetable production and marketing have to be modified. In this regard, there is an urgent need to review the situation of this sector and we have to project a sound plan for its future growth.

While considering these aspects, I felt that, there is the need for a comprehensive book giving details of the vegetable crops of India in terms of pre and post-harvesting aspects. The lack of a up to date book on these aspects made me to take up this task without any delay.

The purpose of this book is to highlight the importance of this forgotten sector in our economy with the help of necessary supportive available data. This will be useful to the policy makers to prepare the future policies, to the researchers, to the present and future citizens of our country and also to the traders who are all engaged in this business.

In this book, I do not pretend to claim credit for any original contribution on the subject. Rather, it should be viewed as a concise collection of various topics from different books, bulletins, research papers, journals, annuals etc., I am indebted to all these contributors, organisations and institutions.

I express my gratitude to my wife Mrs. Poornashri and loving children Shreyaswi, Nithin and Sachin for their encouragement and inspiration during the preparation of the manuscript of this book. I am grateful to my father V.S.N. Bhat for his inspirations. I am also grateful to the management, principal, friends and my colleagues for their cooperation.

Vigneshwara Varmudy

CONTENTS

LIST OF TABLES

1
Introduction

The Green Revolution is one of the greatest successes that the country has observed and the country achieved self-sufficiency and a good degree of stability in food grain production. Even after attaining self-sufficiency in food, the country still faces the challenges like malnutrition which shows that nutrition has not yet been considered as an indicator of social change or for that matter, Economic Development. According to the recent National Family Health Survey, over 53 per cent of the children under four are malnourished and underweight for their age, 52 per cent of the children of this age group are stunned (growth) and about 17 per cent of them are wasted with relatively low weight for their height. The country is currently home to about 40 per cent of malnourished children in the world. These figures exist because more than 85 per cent of pregnant women are anaemic in India. High levels of vitamin and iodine deficiency diseases continue to reduce physical and mental performance. Malnutrition results from a combination of several factors, among these, lower intake of vegetables is one among them. So, the nutritional security in a country like India can be achieved ony when enough vegetables are produced and consumed, as it is universally accepted that these vegetables are cheaper source of proteins, vitamins, carbohydrates, minerals and dieetary fibers etc.

A wide variety of vegetables are grown in India. They are grown right from the sea level to snow line under varied agro-ecological conditions. India, now grows nearly 60 different kinds of vegetables. With the exception of possible half a dozen vegetable crops like brinjal, cucumber, colocasia, ridge gourd, sponge gourd etc., most of crops have been introduced into the country during the various periods before the Christian era to the recent times. India occupies the prime position in the production of most of these vegetable crops.

India is the second largest producer of vegetables in the world. According to the latest estimates of the National Horticultural Board, the total area under vegetable crops in 1997-98 was 5.6 million hectares and the national production stood at 72.83 million tonnes. In terms of area and production of the vegetables, it has been increasing year after year during the post-green revolution period and the country is heading towards another revolution in the vegetables sector. However, in recent years, the demand for nutritious food has been increasing and it is estimated that the realistic demand for vegetables in 2030 will be around 250 million tonnes. This clearly reflects that there would be pressure to produce more at much accelerated pace compared to the past. But this is not an easy task, since, there are several pre- and pot-harvest problems in this much neglected and forgotten sector. As most of the vegetable growers are small and marginal farmers in India, there is an urgent need to tackle the existing problems and to have planned sound vegetable policy for the future.

Considering the importance of this sector in our economy, this book concentrates on various aspects of some of the important vegetables like underground, fruit, cole crops, cucurbits, leafy and legumes.

Second chapter of this book discusses the present status of vegetables in India, where a detailed study is made to to highlight its importance. In the Third chapter, a detailed discussion is made on earth vetetables like potato, cassava, sweet potato, yams, carrot, radish, onion and garlic. Fourth chapter concentrates on fruit vegetables like tomato, brinjal, chilli and okra, while the Fifth chapter discusses on cole crops and the Sixth chapter on cucurbits. The Seventh and Eighth chapters deal on leafy and legume vegetables. All of these chapters give importance to the basic aspects of these vegetables like origin, values and uses, area and production both in the world and in India, varieties grown, harvesting, maketing, problems and prospects.

Chapter Nine deals in detail about the maketing practices of vegetables. Here, stress is given to packaging, transportation, grading, storage, processing, marketing channels, export etc. In the

Tenth chapter, an effort is made to highlight the present problems of this sector and an attempt is also made to note-down some of the important measures to be taken for the future growth. A separate discussion is made on the role of organisations in the development of vegetables in India in the Eleventh chapter, which provides necessary information on pre- and post-harvest technologies undertaken in our country for the development of vegetables over the years.

Vegetables : Present Status

The term vegetables is usually applied to edible plants which store up reserve food in roots, stems, leaves fruits and are eaten raw as *salad* or cooked. The vegetables rank next to cereals as sources of carbohydrate food. The nutritive value of vegetables is tremendous since it has mineral salts and vitamins. Different varieties of vegetables belonging to the tropical, sub-tropical and temperate groups are cultivated in India. Majority of the vegetables, being short-duration crops fit very well in the intensive cropping systems and are capable of yielding high and giving better economic returns. These aspects clearly show that the cultivation and the use of vegtables will definitely improve the living standard of the people in a country like India and thereby it is possible to obtain the ultimate goal of economic development.

Vegetables in Development

Economic development is long term process in which several forces and factors of production work together to bring about an economic change for the betterment. So, as to measure economic development, several criteria have been noted viz., quantitative and qualitative. In recent years, the qualitative aspects are gaining importance for which 'The Human Development Index and Physical Quality of Life Index' are used. Human development is conceived as a process of enlarging human capabilities and choices. Apart from the basic necessities, the human choices include long life, good health and an improvement in the quality of life which includes a reduction in the ratio of poverty or hunger.

In the process of development, India, has observed self-sufficiency in the production of cereals, still the country has failed in the field of nutrition. Because of this, nearly half of all children below five are malnourished, and since their mothers are impoverished too, once-third of all newborns are of low weight. All of these show that the efforts to supply nutritious food in India is still lagging behind. Even though nutritional status of the people is one of the most important indicators of development, the country has failed th provide this basic requirements to its people because of an increase in its population, socio-economic inequity, fragmentation of land and decrease in per capita land area. So as to solve these problems, the available best alternative is growing of vegetables which contribute largely to solve the problems like food and nutrition of the country.

Importance of Vegetable Crops in India

Vegetable Crops play an important role in the developement of our country by improving the economic and social status of the people. The importance of vegetable crops in India can be noted on the basis of the following aspects :

(1) Vegetables are considered protective supplementary food as they contain large quantities of minerals, vitamins and essential amino acids which are required for the normal functioning of human metabolic process. The problems like under-nourishment and hidden hunger can be solved by cultivating and using the vegetables.

(2) Many of the vegetables possess high medicinal value for curing certain diseases.

(3) In sustaining the tempo of agricultural production, diversification is being given high priority, here, the vegetables are considered as an important component of diversification. Thje vegetable crops have tremendous potential of five to eight times more yield increase than cereals and millets.

(4) An increased consumption of vegetables results in qualitative improvement in human diet.

(5) The cultivation of vegetables enhances the income of the farmers and thereby improves the living standards in the rural areas.

(6) Expands the volume and value of agricultural exports.

(7) Creates more job opportunities in rural areas.

(8) It leads to the optimum utilisation of the available natural resources under diversified agrarian system.

(9) Develops the industrial and service sectoral activities.

Nutritive Value of Vegetables

Vegetables play an important role in providing balanced nutrition to the majority of Indian population as these are the valuable source of proteins, minerals, vitamins and to some extent carbohydrates. The nutrients in the vegetables overcome the major disorders like anaemia, deficiency disorders and other ailments in human beings. The Nutritional Expert Group prescribes a minimum of 2,400-3,900 calories of energy, 55g proteins, 0.4-0.5g calcium, 20mg iron as can be observed from Table. 2.1. So as to fulfil these prescriptions, the simple solution available is consumption of more vegetables as they possess, all of these ingredients. Further, for a balanced diet, an adult needs about 2800 g of vegetables per day; of this about 85 g leafy vegetables and 85 g other kinds of vegetables. But the consumption of vegetables in India is much below and the production of vegetables is not even half of the required quantity. So, the scope for an expansion in its area, productiona and productivity, is vast in India.

Table 2.1 : Daily Allowance of Nutrients for Indians

Nutrition	Man	Woman
Calories	2400-3900	1900-3000
Protien (g)	55	45
Calcium (g)	0.4-0.5	0.4-0.5
Iron (mg)	20	20
Vitamin A (mg)	750-3000	750-3000
Thiamine (mg)	1.2-2.0	1.0-1.5
Riboflavin (mg)	1.3-2.2	1.0-1.7
Nicotinic acid (mg)	16-26	13-20
Ascorbic acid (mg)	50	50
Folic acid (mg)	100	100
Vitamin (mg)	1	1
Vitamin D (IV)	200	200

Source : Nutritive Value of Indian Foods, C. Gopalan et.al., 1371, NIN, Hyderabad.

Classification of Vegetables

A good number of vegetables are grown throughout the world, among these nearly fitfty are most popular. These are classified into certain groups. They are : (a) botanical classification, (b) classification based on hardiness; (c) based on the parts used, and (d) based on eassential methods of culture. Classification based on essential methods of culture is the most convenient and is generally followed for describing the cultural operations of different vegetables. The vegetables can be studied under 13 groups, viz., Potato, Solanaceous fruits, cole crops, root crops, bulb crops, peas and bean cucurbits, sweet-potato, okra, *salad* crops, greens, other root crops and perennial vegetables.

Purpose of Vegetable Cultivation

The cultivation of vegetables is undertaken for different purposes. It varies from micro to macro level. In the micro level, it is grown mainly to cater the needs of family while the macro level cultivation is taking place so as to fulfil the needs of the people in cultivation, it can be regarded as home or kitchen gardening and the macro level can be considered as commercial gardening. The commercial cultivation of vegetables is growing in recent years and is becoming popular owing to its food value. In the macro level cultivation of vegetables, it is undertaken on large scale, mainly for the purpose of sale; either in the local market or distant markets. This type of growing is further subdivided into; market gardening, Truck gardening vegetable forcing and vegetable for processing.

World Production of Vegetables

The world production of fresh vegetables in 1999 is estimated at 605536 thousand M.T. which is an improvement over the previous year. The figures given in Table 2.2 show that the production of vegetables has been increasing over the years. As may be noted from the table that China continues to maintain the supremacy in the production of fresh vegetables and its share in the world production of vegetables is 3.74 per cent while in the Asian region its share is 57.9 per cent. It is being followed by India and its share is 9.54 per cent in the global production of fresh vegetables whereas it is 14.25 per cent in the Asian region. The other major producing countries are the USA, Turkey, Italy, Japan, Iran, Spain, Korean Republic and Russian Federation. As a whole, these top ten countries account for about 69.76 per cent of the total production of fresh vegetables in the world.

Table 2.2 : World Production of Fresh Vegetables by Major Producing Countries

Country	1995	1996	1997	1998	1999
China	204465	227138	234616	23416	234616
India	54059	54967	55774	55774	55774
USA	34633	35204	35658	34405	34537
Turkey	21848	22149	21177	21743	21743
Italy	13100	13100	14733	14501	14501
Japan	13636	13683	13518	13565	13565
Iran	8860	9950	12253	13198	13198
Spain	10210	10460	10908	11661	11596
Korean Rep.	11423	10693	10688	10784	11586
Russian Feb.	11680	12220	11796	11298	11298
World total (inluding others)	559933	589133	599546	604685	605536

As a whole, among the fresh vegetables produced in the world, it is being dominated by fresh tomatoes and its share is around 15 per cent in the total. The other major vegetables produced in the world consists of cabbages, dry onions, cucumber and gherkins, watermelons, carrots, chillies, garlic, pumpkins, squash gourds, green peas and roots and tubers. The major consumers of vegetables in the world are China, India, Germany, U.K., U.S.A., France, Italy, Spain and the Gulf nations. In terms of imports, Germany ranks first followed by U.K., Japan, USA, France and others while the major exporting countries are the Netherlands, Spain, USA, Italy, France, Mexico, Belgium and China. The total value of the imports exceeds 25,000 million US dollars and the value of exports in around 2,500 million US dollars. It is interesting to note that, India being the second largest producing co9untry in the world, does not even figure among the ten major exporting countries.

Vegetables Crops in India

Due to the unique geographical and climatic conditions, India is endowed with a wide variety of vegetables. Vegetable crops cultivation has had a chequered history in India and oriental scriptures of the pre-Christian era mention several kinds of vegetables being grown in homestead gradens and pot herbs for their medicinal and nutritive value. Nearly 60 kinds of leafy, fruit and other varieties of vegetables and starchy tubers are being cultivated in India. The status of vegetable production in India has been unique, consisting of diverse kinds as home or kitchen gardens in Urban areas, market gardens such as those near the metropolitan cities and truck farming involving long distance haulage and catering to distant terminal wholesale markets.

In India, a specialised and extensive vegetable growing system like the riverbed growing of cucurbits can be seen along the Ganga, the Yamuna, the Narmada, the Krishna and the Pennar riverine regions. Rainfed cultivation of vegetables is practised in the southern states like Andhra Pradesh, Karnataka, Kerala and Tamil Nadu and on the West Maharashtra or simply one can say that, vegetables are grown in India right from the sea level to snow line under vaired agro-ecological conditions. With the exception of possibly half a dozen vegetable crops like brinjal, cucumber, colocasia, ridge gourd, sponge gourd etc., most of the crops have been introduced into the country during the various periods before the Christian era to the recent times, for example, tomato was introduced about two centuries ago. India is the second largest producer of vegetables in the world next to China. The total area under vegetables in 1999 was 4998 thousand hecrares and the production was 57774 thousand M.T. India's share in the production of vegetables in the world is 9.54 per cent while its share in the Asian region is 14.25 per cent.

India ranks first in the production of cauliflower, pumpkin, squash gourds and green peas in the world, while it occupies second in brinjal, cabbage and onion and third in garlic green beans and watermelon. As far as roots tubers, soya beans, potato, tomato and carrots are concerned, its place is within the top ten in the world. Table 2.3 shows India's share and position in the world production of some of the important vegetables in 1999.

Table 2.3 : India's Share in World Production of Important Vegetables in 1999

Name of the Vegetable	World Production	India's Production	India's share	India's Rank in the World
Brinjal	20147	6000	29.78	2
Cabbage	47574	4200	8.82	2
Carrots	18303	340	1.85	7
Cauliflower	13690	5000	36.52	1
Cucumber and Gherkins	26662	116	0.43	27
Pumpkin, Squash gourds	14664	3300	22.50	1
Tomato	90360	5300	5.86	6
Watermelon	49082	2473	5.03	3
Dry onion	40023	4429	11.06	2
Potato	289780	22100	7.62	5
Garlic	8776	452	5.15	3
Green Peas	6892	2000	29.01	1
Green Beans	4294	400	9.31	3
Soya Beans	157744	6100	3.86	5
Roots & Tubers	626846	29300	4.67	4

Source : FAO

Area and Prduction of Vegetables in India

Vegetables are grown in all the states in the country. In terms of area under vegetables, Orissa ranks first followed by Uttar Pradesh, Bihar, West Bengal, Karnataka, Assam and Kerala. While in terms of production, Uttar Pradesh stands first followed by Bihar, Orissa, Karnataka, West Bengal and Tamil Nadu. Orissa's share in the total area under vegetables in India is 16.43 per cent while that Uttar Pradesh, it is 16.19 per cent and that of Bihar it is 16.06 per cent. The share of West Bengal is 9.56 per cent and that of Karnataka 5.49 per cent. In the production of vegetables, Uttar Pradesh contributes about 19.37 per cent, Bihar 17.15 per cent, Orissa 12.16 per cent, Karnataka 7.96 per cent, West Bengal 7.53 per cent and Tamil Nadu 6.14 per cent towards the total production of vegetables in India. Table 2.4 gives data for state-wise area and production of vegetables in India for the periods 1992-93 to 1995-96.

In terms of the area under some selected vegetables like cauliflower, tomato, okra, cabbage, potato, onion and casava, different states dominate in these. Orissa ranks first in the area under tomato, okra, cabbage and second in cauliflower, while Bihar ranks first in cauliflower and second in tomato, okra, cabbage and potato. Maharashtra stands first in onion and fifth in cauliflower. Kerala is having largest are under cassava and ginger.

Table 2.4 : State-wise Area and Output of Vegetables in India
(Area, Lakh Hectares; Output lakh M.T.)

State	1992-93		1993-94		1994-95		1995-96	
	Area	Output	Area	Output	Area	Output	Area	Output
Andhra Pradesh	1.49	13.68	1.54	14.20	1.54	23.41	1.54	24.44
A.P.	0.17	0.80	0.17	0.80	0.17	0.80	0.17	0.80
Assam	1.52	17.54	2.06	19.32	1.88	19.70	2.49	24.85
Bihar	8.87	162.00	9.14	136.10	8.08	117.26	8.57	122.85
Delhi	0.47	7.35	0.25	4.63	0.51	5.51	0.56	6.13
Goa	0.07	0.60	0.08	0.65	0.08	0.65	0.08	0.65
Gujarat	1.42	15.57	1.15	18.70	1.22	17.30	1.68	20.89
Haryana	0.74	10.30	0.75	11.55	0.85	12.75	0.94	14.20
Himachal Pradesh	0.39	4.76	0.38	5.38	0.38	5.44	0.39	5.69
Jammu & Kashmir	1.80	7.45	0.33	3.54	0.33	3.54	0.33	3.54
Karnataka	4.17	46.62	2.61	50.35	2.90	56.68	2.93	57.05
Kerala	2.03	29.03	2.44	27.90	2.44	27.90	2.44	27.90
Madhya Pradesh	2.14	20.47	1.81	25.51	1.81	22.41	1.90	23.53
Maharashtra	2.01	35.71	2.25	27.38	2.21	28.09	2.16	29.57
Manipur	0.05	0.36	0.04	0.33	0.05	0.35	0.05	0.36
Meghalaya	0.25	2.38	0.25	2.38	0.25	2.00	0.22	2.11
Mizoram	0.12	7.01	0.06	0.45	0.12	0.77	0.12	0.79
Nagaland	0.08	0.67	0.09	1.08	0.08	0.87	0.08	0.87
Orissa	7.60	77.45	7.75	79.83	8.01	78.96	8.77	87.06
Punjab	0.85	14.55	1.03	17.21	1.03	17.21	1.05	17.74
Rajasthan	0.60	2.02	0.67	3.63	0.68	2.83	0.76	3.57
Sikkim	0.08	0.52	0.08	0.47	0.09	0.29	0.0	0.51
Tamil Nadu	8.54	39.59	1.71	43.89	1.75	43.98	1.75	43.98
Tripura	0.32	3.21	0.32	32.09	0.32	32.09	0.32	35.85
Uttar Pradesh (Hills)	0.54	9.59	0.84	7.18	0.87	7.74	0.88	7.91
Uttar Pradesh (Plains)	5.93	89.11	6.24	103.60	7.43	119.12	7.76	130.83
West Bengal	6.10	129.77	4.70	48.59	4.90	53.40	5.10	53.91
A&N Islands	0.04	0.15	0.03	0.19	0.03	0.16	0.03	0.16
Dadra & Nagar Havelli	0.02	0.14	0.02	0.14	0.02	0.14	0.02	0.14
Pondicherry	0.02	0.23	0.02	0.23	0.03	0.37	0.02	0.34
All India	50.45	638.06	48.76	657.87	50.13	672.86	53.35	715.94

Source : National Horticultural Board.

As a whole, the total area under vegetables in India has been increasing slowly and steadily as can be observed from Table 2.5 where the FAO estimates are given. However, the estimates made by the National Horticultural Board show that the total area under vegetables has been fluctuating over the years. According to its estimates, the total area under vegetables in India in 1991-92 was 55.93 lakh hectares and from then onwards it came downward and in 1995-96 it was 53.35 while in terms of production, it shows an upward trend. In 1991-92, the production of vegetables was 585.32 lakh M.T. and it went up to 1106.20 lakh M.T. in 1997-98. Based on the nutritional requirment for the current population of 1000 million, the NCA has fixed the required volume of vegetables at 104 million tonnes which is far above the present production estimates for FAO. Hence, the scope for an expansion in area and to increase the productivity and production is always there in our country.

Table 2.5 : Area and Production of Vegetatbles in India

Year	Area	Production
1990	4799.95	48936.58
1991	4864.04	49967.24
1992	4533.05	50967.87
1993	4768.63	52250.80
1994	4789.90	52647.40
1995	4901.90	54059.30
1996	4998.00	54967.00
1997	4998.00	55774.00
1998	4998.00	55774.00
1999	4998.00	55774.00

Source : FAO QBS Vol. 12 1999.

HYBRIDS IN VEGETABLE CROPS

The productivity per hectare and the quality of vegetables actually determines the fate of the growers in particular and that of the nation in general. For realising higher income from an Unit area and to improve the processing and exports, suitable high yielding varieties are essential. In this regard, efforts were made since the 70s in our country, however, it has started improving in the recent years. As the scope for extension in area under vegetables in India is limited, the production can be increased only with the help of F1 hybrids of various vegetables.

Meaning of Hybrid

In some crops, from the union of different gametes, there appears a developemental stimulers signals in the progeny called F1 generation greater vitality, rapid growth and development, higher productivity, resistance adaptation and uniformity. The expressions singly or collectively indicate hybrid vigour and the product–a hybrid or F1. These hybrids in general offer advantages of early maturity and better keeping quality.

In India, the first report of hybrid vigour in chilli was made at IARI in 1933. However, the first hybrid was developed by Indian Scientists for bottlegourd in 1971, followed by the development of F1 bybrids for summer squash, cucumber and brinjal. Since then, several hybrids of vegetables have been developed in various research institutions and released for cultivation under the All India Co-ordinated Vegetable Improvement Project. So far, 131 open pollinated varieties, 36 hybrids, 3 synthetics and 29 resistant varieties covering 20 different vegetable crops have been identified as promising for release in different agro-climatic regions. Of these 199 releases by the AICRPV, 13 are in cauliflower, 9 in muskmelons, 40 in tomato, 45 in brinjal, 12 in chillies, 20 in pea, 16 in onion and 44 in other vegetable crops.

Work on the development on F1 bybrids of various vegetable crops has been carried out at IARI and its research centres. Apart for these private sector, seed companies are also playing an important role in developing F1 hybrid seeds in India. So far these private companies have released 11 hybrids for brinjals, 9 for tomatoes, 8 for watermelons, 4 for capsicum, 3 for cabbage, 2 for muskemelon and one each for carrot and chilli. Apart from these, the country is also importing F1 hybrid seeds since 1988, under OGL.

Because of the introduction of these seeds, the total are under F1 hybrids in the vegetables in India has expanded over these years and the productivity has shown an upward trend especially in tomato and cabbage. At present, the share of hybrid vegetables in their respective total cropped area ranges around 31.51 per cent in tomato, 31.34 per cent in cabbage, 17.80 per cent in brinjal, 5.38 in okra, 4.02 per cent in melons, 3.29 per cent in cauliflower, 2.44

per cent each in chillies and gourds respectively. Through hybrids, the average productivity has increased in tomato by 49.7 per cent, 45.3 per cent in brinjal, 53.8 per cent in capsicum, 52.7 per cent in green chillies, 55.7 per cent in cabbage and 49.4 per cent in cauliflower.

The above aspects clearly show that the hybrid technology is going to stay in our country and it forms an important component of our plans for increasing the production of vegetables. However, there is the need to improve the prevailing situations in this so as to gain the real benefits. So, what is required now is cost effective technology to promote a wider spread, like multiple disease resistant hybrids. Resistance breeding should be integrated with hybrid technology to increase productivity. Along with these, there is the need for a national seed policy to promote healthy developemtn of both public and private sectors. Efforts are also required to supply these seeds at a lower price. As the country is blessed with agro-climatic conditions suitable for year round production of all types of vegetables, it calls for appropriate research efforts to envolve suitable hybrid varieties which can meet the domestic requirements both for consumption and processing and also for exports.

3
Underground Vegetables

The underground vegetables or earth vegetables include all forms in which food is stored in underground parts. The storage organs may be quite different morphologically. Some of them are true roots, others represent modified stems like rootstalks, tubers and bulbs. These vegetables have been playing an important role in the development of the economies over the years. These are extremely valuable since they are readily digested and have high energy content. In almost all countries, these vegetables are cultivated both for human consumption and for stock feed.

The various earth vegetables will be grouped according to their morphological origin. Of these, some of the important vegetables are taken into consideration in this chapter which includes potato, cassava, sweet potato, yams, carrots, radishes, onion and garlic. Among these, India has got an important position in terms of production in the case of potato, onion, garlic etc., all of these vegetables play an important role in the living conditions of the small and marginal farmer in our country. However, there are several problems in these both in terms of production as well as in the post-production aspects. The improvement and development of these underground vegetables are essentially required in our country since they are also considered as the valued food products and they supply the much needed nutrition to the people.

POTATO

Potato (*Solanum tuberosum*) is one of the most important food plants of the world. It is a native of South America and was cultivated from Chile to New Granada at the time of the Spanish explorers. The first mention of the potato in literature was in 1553, and first published illustration appeared in 1633. It was introduced into Europe soon after 1580 by the Spaniards and by the end of the 17th century, it had spread all over Europe.

The potato was brought to India in the 1600s by the Portuguese traders, who landed north of Bombay. Prior to 1700, it was grown as a garden vegetable in parts of Western India, while it reached Southern India only in the 1880s. The British promoted potato cultivation in the hills and it spread to the plains and by 1900, small plots of potato were found near towns scattered throughout India.

Potato is an errect, branching, more or less spreading annual from 2 to 3ft in height. It has pinnately compound leaves, fine fibrous roots and numerous rhizomes which are swollen at the tip of form the familiar tubers. The flowers are white, yellow or purple with a tubular corolla, while the fruit is brownish-green or purple inedible berry. Potatoes are adopted to many soils and many climates. They are, in fact, grown the world over, except in low tropical regions. The best environment is a cool moist climate.

Values and Uses

Potatoes contain about 78 per cent water, 18 per cent carbohydrates, including a little sugar as well as starch, 2 per cent proteins, 0.1 per cent fat and 1 per cent potash. Potato is a nutritious food. Apart from providing carbohydrates, proteins, vitamin C and a number of Vitamins of the B group, it provides high quality dietary fibre. The protein is comparable in quality to egg and milk proteins and is superior to cereal and other vegetable proteins. It also produces more protein than all the other major food crops per unit area and time. For instance, the potato produces 3 kg of protein/ha/day as compared to only 2.5 kg/ha/day in wheat, 1.2 kg/ha/day in meize and 1 kg/ha/day in rice.

The tubers are used as vegetable. In European countries, they make a universal table food. Small tubers are utilized for the production of starch and industrial alcohol. Potatoes are also fed to livestock. Fresh potatoes are used in making potato chips and then used after frying. Among the processed potato products, chips and French fries are the most popular forms in different countries. In India, chips are very popular in the urban areas, while French fries are not commercially available in the market. They are served in the restaurants or at fast food outlets and are sometimes prepared fresh at home. The annual per capita consumption in our country is 15 kg as against the world average consumption of

24 kg. In European countries, it is in the range of 80-155 kg per year on fresh weight basis alone. As a whole, the per capita consumption of potatoes in India has gone up from 12 kg in 1960 to 40 g in 1997 registering over 233 per cent increase.

Production of Potato in the World

In terms of production of potatoes, China ranks first followed by Russia, Poland, U.S.A. and India in the world. The total production during 1999 was 289.78 million tonnes. The average yield in India is significantly higher than China and Russia. As a whole, the trend in production in the world over the years shows ups and downs as can be seen from Table 3.1

Table 3.1 : Production of Potato in Major Producing Countries in the World (in million tonnes)

Country	1994	1995	1996	1997	1998	1999
China	43.84	.45.75	52.03	57.31	44.47	43.47
Russia	33.83	39.90	38.70	37.04	31.41	31.41
Poland	23.06	24.89	27.22	20.78	25.94	26.00
USA	21.19	20.12	22.62	21.11	21.58	21.40
India	17.39	17.40	19.24	25.06	19.60	23.00
Ukraine	16.10	14.73	18.41	16.70	15.40	15.40
Germany	10.64	10.89	14.26	12.06	11.71	12.07
Belarus	8.24	9.50	10.88	6.94	10.00	10.00
World Total (including others)	270.54	285.14	311.18	301.46	282.10	289.78

Source : FAO

Area, Production and Producitivity of Potato in India

In India, potato is cultivated in almost all states. Uttar Pradesh, West Bengal and Bihar accounts for nearly 75 per cent of the area under the crop and about 82 per cent of the total production. In terms of production, Uttar Pradesh accounts for about 40 per cent of the total in India, and in this state Farrukabad and Kannauj are the two main potato growing regions where the productivity is also maximum. While West Bengal's share in Indian total production of potato is about 25 per cent. In West Bengal, potato is cultivated mainly in the districts of Howrah, Hooghly and Burdwan and to a small extent in other districts also.

In Maharashtra, potato is cultivated in Pune, Satara and Sangli whil in Karnataka the Kharif potato is grown in Hassan, Chickmagalur, Belgaum and Dharwad districts and the Rabi in Kolar and Bangalore districts. In Haryana, it is grown in Panipat, Yamunanagar, Karnal, Kurukshetra, Pipli and Shahbad and in Punjab in Amritsar, Ludhiana, Hoshiarpur, Moga, Jalandhar, Patiala and Bhatinda. As a whole, the total area under Potato in 1998-99 was 12.35 lakh hectares and the production was 235.62 lakh tonnes. The productivity of potato was 19.08 tonnes per hectare during the same period. Gujarat stands first in terms of productivity followed by West Bengal, Uttar Pradesh, Haryana and others as can be seen from Table 3.2.

Table 3.2 : State-wise Area, Production and Productivity of Potato in India
(Area in lakh hectares, Production in lakh M.T. and Productivity MT/hectare)

State	1997-98			1998-99		
	Area	Production	Productivity	Area	Production	Productivity
Bihar	2.90	29.35	9.09	3.00	40.00	13.33
Gujarat	0.28	7.10	23.35	0.32	7.88	24.63
Haryana	0.14	1.50	10.71	0.16	2.50	15.63
Madhya Pradesh	0.47	5.47	11.64	0.55	8.10	14.72
Punjab	0.50	10.00	20.00	0.50	7.25	14.50
Uttar Pradesh	4.00	60.00	15.00	4.10	94.00	22.93
West Bengal	2.60	65.00	25.00	2.45	60.00	24.49
Total (Including others)	12.04	192.95	16.03	12.35	235.62	19.08

Source : Ministry of Agriculture.

In general, in terms of area, production and productivity, it has been increasing in India over the years as can be observed from Table 3.3. Since 1950, the annually compounded growth rates of production, area and yields of potato in India have been respectively 6, 3.5 and 2.41 per cent. In terms of production, area and yield, it increased by 12.5, 4.9 and 2.6 times respectively. In terms of its contribution towards the total value of agricultural output of the country, it was 1.4 per cent in 1970-71 and by 1995-96, it increased to 1.81 per cent.

Table 3.3 : Area, Production and Productivity of Potatoes in India
(Area in lakh hectares, Production in lakh M.T. and Productivity in M.T./hectare)

Year	Area	Production	Productivity
1950-51	2.40	16.60.	6.92
1955-56	2.80	18.59	6.64
1960-61	3.75	27.19	7.25
1965-66	4.79	40.79	8.51
1970-71	4.82	48.07	9.98
1975-76	6.22	73.06	11.74
1980-81	7.29	96.68	13.26
1985-86	8.43	104.23	12.36
1990-91	9.42	152.54	16.19
1991-92	10.30	163.88	15.90
1992-93	10.75	157.18	14.62
1993-94	10.80	180.40	16.72
1994-95	10.69	174.01	16.27
1995-96	11.10	188.40	16.93
1996-97	11.10	242.00	19.18
1997-98	12.04	192.95	16.03
1998-99	12.35	235.62	19.08

Source : D of E&S.

Varieties

There are many varieties of potatoes grown in India. They are either introduced into or bred in India. The CPRI has released 34 varieties for different ecological zones of India. They are :

(1) *For South Indian hills :* The varieties recommened for this are Kufri Neela, Kufri Jyothi, Kufri Neelamani, Kufri Muthu and Kufri Swarna.

(2) *North Indian hills :* For this, the suiable varieties are Kufri Kumar, Kufri Kudan, Kufri Jeevan, Kufri Jyoti and Kufri Giriraj.

(3) *Plateau region :* Kufri Kubu, Kufri Chandramukhi, Kufri Lauvkan, Kufri Badasha, Kufri Jawahar and Kufri Pukhraj.

(4) *North Indian Plains :* Kufri Kisan, Kufri Kuber, Kufri Safed, Kufri Sindhuri, Kufri Alankar, Kufri Chamatkar, Kufri Chandramukhi, Kufri Jyoti, Kufri Dewa, Kufri Badshah, Kufri Bahar, Kufri Lalima, Kufri Kawahar, Kufri

Sutlej, Kufri Ashoka, Kufri Pukhraj, Kufri Chipsona-1, Kufri Chipsona-2 and Kufri Anand.

(5) *North Eastern Plains and Hills* : Kufri Red, Kufri Khasigaro, Kufri Naveen.

Harvesting

The potato crop shows the sign of maturity in about three months after planting. The leaves turn yellow, and are shed in course of time and haulms dry up and die which may be taken to be an indication of maturity. However, the harvesting time is determined by the prevailing market price and crop to follow after Potato. Sometime, they are harvested when the tubers are immature in order to get a higher price in the market. After harvesting, they are graded as seed size tubers, large size tubers and chats.

Potato Storage

Over the years, there appeared an increase in the production of potato in India and has led to several post-production problems and the major problem is that of storage. As most of the potatoes are harvested in February-March, it results in a situation where, in times of glut, the price fetched is so low that it is not sufficient even to cover the cost of production. In this regard, storage helps in regulating the supplies to the market and thus avoiding gluts and distress sale by growers. Apart from these, it reduces transport bottlenecks at the peak period of production.

Nearly 90 per cent of the products are produced in the indo-gangetic plains and this region grows only one crop in a year and is harvested in Feb-March. As a result of this, it is available only for 2-3 months in a year, so far meeting the demand of the remaining months, it requires proper storage facilities. Again, there is the need to preserve the seed potatoes upto October which also calls for storing facilities. Hence, cold storage is an essetnial requirement.

In India, potatoes are stored in cold store at 2-4°C. At present, there are 3443 cold store units in our country, of which 2,975 are in the private sector and 303 in the co-operative sector and the rest in the public sector. The present cold storage capacity in the country is about 10.3 million tonnes and it has been estimated that an additional 1.2 million tonnes cold storage capacity will be required in the next five years.

As the number of cold storage plants are limited, most of the farmers in India use the traditional methods of storing and store the table potatoes for a few months. The methods of storage are region-specific. In Karnataka, Maharashtra, Gujarat and Uttar Pradesh, heaps are popular. In Madhya Pradesh heaps and pits are commonly used for potato storage, while in Bihar they are stored in machine. The efficiency of these are varying and the scope for losses are very high still they are popular because they have several advantages like (a) they are cheap; (b) the materials required are locally available and (c) acceptance of the farmers. Again, the high cost of cold-storage reduces the scope for adopting it by the farmers, which makes the way for opting the traditional methods of storing in India.

Marketing

There is no organised market for potato in India. Several studies on the potato marketing in different states showed that it is oligopolistic, as the role of commission agents is maximum. In most of the cases, the producer's share in consumer's rupee varies in between 51-75 per cent. Normally, the farmer's share depends on the marketing channel and the costs in marketing the potato by the producers. The number of marketing channels varies from place to place, however several studies show that there are more than 10 channels in Uttar Pradesh. The share of the producers where highest when the potato was sold directly to the secondary market from cold storage and the lowest when sold to the village trader. At all India level, the Directorate of Marketing and Inspection (DMI, 1984) showed that producers net average share was 59 per cent of the consumer's price. Marketing costs averaged about 19 per cent and the marketing margin 22 per cent of the consumer price. In the case of Bihar, the producer's share was reported only 42 per cent of the consumer price. The price spread was high due to (a) Poor transport facilities in rural areas, (b) Absence of grading; (c) Lack of adequate storage facilities; (d) Long chain of middleman; (e) High assembling charges and (f) Malpractices in the markets.

In Karnataka non-availability of the basic facilities like storing, grading, loading and unloading in the market yards leads to heavy losses to the farmers. As a whole, the prevailing marketing system is segmented and the farmers are not getting the real benefit which is due to them.

Price Behaviour

During the agriculture year (July-June), the wholesale prices of potato ruled high till August 1997. The index number of wholesale prices (1981-82 as the base) stood at 280.9 in July 1997, increased to 303.4 in August, 1997 showing an increase of 8.0 per cent over the index in July 1997. Since September 1997, the index gradually declined and continued up to Jan. 1998, from Feb. 1998 again started increasing and reached 791.5 in June 1998. At this level, the index number of wholesale prices of potato was higher by 51.8 per cent over the price index in July 1997. The annual average wholesale price index during 1997-98 was 2.6 per cent lower over the corresponding index annual averae of 1996-97. Table 3.4 gives the index number of wholesale prices of potato for the year 1997-98 along with the corresponding price index during 1996-97 and percentage variation over the year.

Table 3.4 : Index Number of Wholesale Prices of Potato

Month	1997-98	1996-97	Percentage variation over the year
July	280.9	521.3	− 46.1
August	303.4	538.9	− 43.7
September	265.3	543.3	− 51.2
October	263.7	584.3	− 54.9
November	272.3	600.3	− 54.6
December	248.8	530.7	− 53.1
January	239.0	403.7	− 40.8
February	281.7	302.4	− 6.8
March	419.1	278.2	+ 50.6
April	628.3	221.3	+ 183.9
May	746.9	167.6	+ 345.6
June	791.5	176.5	+ 348.4

As potato is a bulky and perishable product, lack of storage and marketing facilities and even a marginal increase of production can cause a change in the price level. It has been experienced that every 3 to 5 years when there is an increase in potato production by 20 to 50 per cent, a glut situation arises and potato price crashes. Apart from these, non-availability of basic facilities in the market yards results in drastic fall in prices.

Price Policy for Potato

To protect the interests of the farmers, the Government fixes the minimum support price for major agriculture products. However, in the absence of minimum support price and to protect the interest of potato cultivators, the Government has bought potato under Market Intervention Scheme when the price of potato fall to uneconomic level. The decision to launch Market Intervention Scheme is taken by the Central Government on a special request from the State Governments for a particular period for a fixed quantity at a predetermined price. The NAFED along with state designated aencies makes, purchases and shares losses if any, on 50 : 50 basis.

Processed Products

Potato processing is only in the initial stage in India while in the developed countries, they are popular. Processing of potatoes is an essential requirement in our country so as to avoid gluts and the consequent difficulty of storing large quantities of potatoes during the period of extremely high temperature.

The processed potato products can be classified as follows :

(1) Fried products like potato chips, frozen French fries and other frozen products.

(2) Dehydrated products, such as dehydrated chips, dices, flakes, granules, flour, starch, potato custard powder, soup or gravy thickener and potato biscuits.

(3) Canned potatoes.

In India, many processing plants have been installed and the capacity of potato chips in the organised sector was 6,000 tonnes during 1990 adn several brands of chips are available in the market. In India, processing of potatoes constitutes less than 0.5 per cent of the total annual production. All of these indicate that the scope for processing is vast but India lags behind in it. As the demand for easy-to preparea and fast foods are growing, there is a good scope for processed potato products in our country. The prevailing size of domestic market for potato chips is about 20,000 tonnes of which 16,000 tonnes is being held by local producers. The market for branded chips value is about Rs. 20 crores. In the branded chips, the leaders are Uncle Chips, Fri to Lays and SM Dyechem.

Exports

In the international market for potatoes, India's share is more or less negligible. This is due to high price and non-uniformity in the quality of its produce. In 1993-94, India exported 15,755 tonnes of potatoes valued at Rs. 6.69 crores and the maximum of 34,516 tonnes valued at Rs. 18.90 crores was exported in 1995-96, however, from then, it came downwards and can be seen from Table 3.5. India exports potatoes mainly to Mauritius, Nepal, Sri Lanka, Turkey and UAE.

Table 3.5 : Export of Potato from India

Year	Volume (M.T.)	Value (Rs. Crores)
1980-81	7,219	1.18
1985-86	1,970	0.37
1990-91	2,496	0.74
1991-92	4.692	0.87
1992-93	5,661	1.20
1993-94	7,093	2.13
1994-95	15,755	6.69
1995-96	34,516	18.90
1996-97	24,936	17.16
1997-98	20,884	9.04

Source : DGCIS.

Problems

(1) Non-availability of certified seed to the farmers in many parts of the country.

(2) Absence of sustainable technologies for both pre and post-harvest operators.

(3) As the water requirement of this crop is more, water scarcity has become a major problem.

(4) Fluctutation in the yields in different parts of the country; this appears to be there because of inadequate diffusion of improved production technology.

(5) *Marketing problems.* As potato is a semi-perishable and bulky product, inadequacies and malpractices in cold storages, transport bottlenecks and a poor marketing system reduces returns to the growers and increases cost to the consumer.

(6) Non-availability of processing varieties of potato. As the demand for processed potatoes is increasing, there is the need for such varieties which are suitable for it, but the available varieties are insufficient.

(7) Non-availability of storage facilities in accordance with the purpose for which potatoes are stored.

(8) Increasing level of production cost because of the high price of seed and other inputs. This results in high cost of raw materials for the processing industry and on the other hand, the domestic prices of potato are relatively high as compared to international prices, making the export of potatoes and pottao-based products only modertely competitive.

Prospects

Potato is a crop that can be grown from the sea level to the snow line and has a wide flexibility in its planting and harvest time and can fit in well with various intensive cropping systems including inter-cropping systems. Thus, there is a great opportunity to increase the area and production of potato. Apart from these, its high nutritional value, culinary adaptability, processibility and agronomic adaptability and efficiency, there is a strong case for promoting the diversified use of potato and raising the status of staple food. As the population of the country is growing at a faster rate and the present per capita consumption of potato is much lower, there is a vast scope for an improvement and development of this sector in India. By considering all these aspects, what is needed now is proper solutions for the prevailing constraints. In this regard, the following measures will be useful, they are :

(1) There is the need to bring more area under potato by involving potato as an intercrop. Along with this, there is also the need to expand its cultivation to non-traditional areas through the use of high thermoperiod resistant varieties. By developing suitable production technologies and disease control methods, it is possible to exploit the potential of the kharif crop also.

(2) There is the need to develop varieties having durable resistance to diseases and pests and capable of giving high yield with low levels of inputs and evolve environment-friedndly agrotechniques.

(3) Emphasis should be there to recognise and supply varieties suitable for processing.

(4) Work has to be carried on socio-economic constraint analysis in adoption of new potato production technology, technology assessment and refinement through proper linkages and analysis and forecasting of potato prices.

(5) So, as to reduce the dependence on energy intensive refrigerated strogage and to avoid sweetenting and the consequent deterioration of tuber quality that occurs under refrigeration, there is a need to develop varieties with good keeping quality.

(6) There is an urgent need to formulate a sound policy for the storage of potatoes in India.

(7) There is the need to utilise larger volume of potatoes in the processing industries. In this regard, appropriate processing and storage technology, proper packaging and transportation facilities and creating awareness among the public etc., will be useful. Along with these, there is the need to improve and enhance the efficiency of processing and to reduce the cost of processed products. Development of appropriate technology for dehydrated products for the rural sector will not only ensure remunerative returns to the small farmers but also boost the processing industry in India.

(8)   So as to avoid the problem of marketing, there is the need to start potato grower's cooperatives in the production areas.

(9)   So as to avoid glut situations and price crash of potatoes, the major potato growing states should resort to advance forecasting of area under potato and work out strategies to divert surplus potatoes either to deficit areas or by way of exports. NAFED and APEDA have to work fully in respect of potato marketing and export.

(10) To make the export venture a success, the governement and other exporting agencies should procure potato from the major production centres at supportive price for exports. India has the potential of becoming a major exporter of both ware and seed potatoes. More than 80 per cent of the crop is grown in the winter when there is no potato crop in the temperate countries. In this regard, there is the need for a long term policy.

CASSAVA

Cassava (*Manihot esculenta*) is one of the most important of the tropical root crops and is a basic food for millins of people. It is a native of South America and widely grown in all tropical and sub-tropical regions. The Cassavas are shrubby perennials with stems that reach a height of above 9 feet. It is also called manioc, mandioc, or yuea, is one of the wholesome foods.

Values and Uses

The Cassava root contains 30-40 per cent dry matter, which is principally carbohydrate. It has acceptable levels of a B Vitamin and provides other minerals too. The leaves contains high levels of protein i.e., 8 to 10 per cent on fresh weight and various parts of Cassava plant also have medicinal value. As human food, the Cassava root is prepared in many ways viz., boiled, baked, fried as meal, flour and even as beet. Starch extracted from the root is used to make a wide range of sweet and savory foods such as tapica pearls, noodles and cheese breads.

In the world, two-thirds of the total production of Cassava is used as human food. In Africa, about 90 per cent of the total produce is used for direct human consumption. In Latin America too, Cassava is basically used for direct human consumption. A number of fermented food products are also prepared in Asia, Africa and Latin America from Cassava. Baked slices known as Kripik are popular in Java; gari is widely consumed in West Africa. Until the fifties when rice was in short supply, almost the entire Cassava produced in India was used for direct consumption. However, in recent years, its direct consumption came down steadily with the increasing availability of cereals. Now, half of the products are used for industrial purposes making edible products such as sago-perals and liquid glucose and a part is used in stock feed. Starch is also manufactured for paper, laundry and other industrial products. Raw Cassava starch is used for healing purposes, and is fermented into an intoxicating beverage.

Cassava is grown extensively in the tropics of Africa and Asia. The major producers are Nigeria, Brazil, Thailand, India, China, Indonesia and Zambia. The total area under Cassava in the world is 16.57 million hectares with an output of 165.46 million tonnes of tubers. Nigeria occupies the first position in area under Cassava accounting for 16.5 per cent of the world area and producing 18.5 per cent of world Cassava followed by Brazil, Cango, Thailand and Indonesia. As a whole, these countries constitute 50 per cent of the area under Cassava (Table 3.6)

Table 3.6 : Production of Cassava in Major Producing Countries in the World ('000 mt)

Country	1980-91	1995	1996	1997	1998	1999
Nigeria	20817	31404	31418	30400	32695	32695
Congo Demep. R.	18694	19378	16800	16800	17060	17100
Brazil	24144	25423	24584	24354	19385	20172
Tanzania	7383	5969	5992	6444	6128	6128
Indonesia	16300	15442	17003	16103	14728	14728
Thailand	21557	17388	18084	18000	15591	16930
Philippines	1840	1957	1900	1900	1787	1787
Vietnam	2439	2212	2067	2067	1783	1783
India	5070	5929	5979	5979	6000	6000
China	3282	3501	3501	3501	3651	3651
Paraguay	3371	3054	2649	3155	3300	3500
Uganda	3406	2224	2245	2291	3204	3400
Ghana	3215	6612	7111	6800	7227	7227
Angola	1613	2400	2400	2326	3211	3211
World Total (Inclu-ding others)	154585	165436	165650	164741	162249	165469

Source : Up to 1997 FAO Vol. 51 & for 1998 and 1999. FAO QB 5 VOL. 12.

Cassava in India

In India, the cultivation of Cassava is mainly undertaken in Kerala, Tamil Nadu, Andhra Pradesh, Meghalaya, Assam etc. Kerala stands first in terms of area under Cassava in India which is about 55 per cent of the total and in terms of production, it stands next to Tamil Nadu while Tamil Nadu ranks first in production and productivity in India and second in terms of area as can be observed from Table 3.7.

Table 3.7 : State-wise Area, Production and Productivity of Cassava in India (1996-97)

State	Area ('000ha)	Production (000 tonnes)	Productivity (Tonnes/ha)
Andhra Pradesh	22.0	174.50	7.93
Assam	2.40	11.50	4.79
Karnataka	0.90	7.10	7.89
Kerala	142.00	2558.30	18.23
Meghalaya	3.90	21.50	5.51
Tamil Nadu	65.70	3043.20	46.32
Total (including others)	260.80	5868.30	22.50

Source : D of E&S.

As a whole, in India the cultivation of Cassava is undertaken in an area of 0.25 million hectares and its production is around 6 million tonnes. It is interesting to note that though India does not have a major area under this crop, its productivity is the highest in the world which is 24 tonnes per hectare. The trend in area and production over to years shows several ups and downs, however, in terms of productivity, it has moving positively as can be seen from Table 3.8.

Cassava became a popular cereal substitute towards the end of the 19th century in our country. It covers seven per cent of the net cropped area in Kerala. In Tamil Nadu, it is grown in the districts of Salem, Kanyakumari, South Arcot, Tiruchi and Dharmapuri. In Andhra Pradesh, Cassava farms are concentrated in East Godawari and Sreekakulam districts.

In Kerala, about 40 per cent of Cassava is raised as mixed crop. In Tamil Nadu and Andhra Pradesh, Cassava is grown under open conditions. However, about 40 per cent of Cassava in Tamil Nadu is intercropped with short duration crops such as groundnut, cowpea, black gram, onion and vegetables. In

Karnataka, it is grown along with areca, coconut and rubber gardens. The mixed cropping system practised in these states provides the much needed additional income to the small farmers.

Table 3.8 : Area, Production and Productivity of Cassava In India

Year	Area ('000ha)	Production (000 tonnes)	Productivity (Tonnes/ha)
1967-68	347.10	4643.70	13.38
1970-71	345.00	5130.00	14.87
1975-76	392.00	6638.00	16.93
1980-81	321.00	5867.90	18.28
1985-86	276.00	4884.00	17.70
1990-91	243.00	5111.00	21.03
1991-92	250.90	5832.50	23.25
1992-93	234.90	5412.80	23.04
1993-94	245.80	6028.90	24.53
1994-95	242.80	5929.30	23.42
1995-96	228.20	5443.20	23.85
1996-97	260.80	5868.30	22.50
1997-98	244.00	5979.00	24.50
1998-99	250.00	6000.00	24.00

Source : D of E&S.

Varieties

The availabe hybrid Cassava vareties are namely H-97, H-165 and H-226, but these did not find much acceptance among the farmers because of the poor quality of Cassava hybrids as compared to the locally popular cultivars and also quality aspects of the product. Most of the available varieties are by duration.

Harvesting

The tubers become ready for harvest in eight to twelve months, depending upon the variety. For domestic consumption, it is harvested in about six months after planting.

Marketing

There is no regulated marketing system for Cassava. The farmers either themselves market it or sell it through contract system. In the contract system, the contractor strikes a bargain with the farmer and a price is fixed. Beacuse of this, the effective price received is always less than what he gets by undertaking to retail himself.

Cassava based Agro Industries in India

Cassava is used as a raw material for a number of value added industrial products such as starch, sago, liquid glucose, dextrin etc. The number of starch and sago factories using Cassava had increased from mere 51 in 1955 to more than 1000 at present. Tamil Nadu holds the key position as the largest producer of starch and sago and most of these industries are located in and around Salem district. As a whole, there are more than 430 Cassava starch industries in Tamil Nadu and more than 500 sago and sago products industries in this state. As far as Cassava chips, flour and sago processed products industries are concerned, the number in the state of Tamil Nadu is more than 100. At present these industries are meeting about 80 per cent of demand of the Cassava based food products of the country besides offereing employment for over 5 lakh people in rural areas. Apart from these industries in Tamil Nadu, the sago and other industries are also functioning in Andhra Pradesh and Kerala too.

Production, Marketing and Consumption of Cassava and its Value Added Products in India

(1) *Sago* : It includes both *motidana* and medium *dana* which are produced in Andhra Pradesh and Tamil Nadu. These are produced for human consumption. The marketing centres for these are located in West Bengal, Maharashtra, Utter Pradesh, Assam, Tripura, Andhra Pradesh, Tamil Nadu and Kerala.

(2) *Nylon Sago* : It is produced in Tamil Nadu for human consumption. The marketing centres for this are Maharashtra, Tamil Nadu and Andhra Pradesh.

(3) *Broken Sago* : It is used in the Textiles and sizing industries and the production of this is taking place in Andhra Pradesh and Tamil Nadu. The marketing centres for this West Bengal and Maharashtra.

(4) *Starch* : Starch is produced in Tamil Nadu and Andhra Pradesh. It is used in textile industreis, Adhesives, Dextrin, Pharmaceuticals, confectioneries, laundry,

foundry and paper industries. Starch is marketed in Maharashtra, West Bengal, Gujarat, Tamil Nadu and Andhra Pradesh.

(5)   *Chips and Flour* : Andhra Pradesh, Tamil and Kerala states are involved in the production of chips and flour. These are used in cattle feed mix plants, adhesives, sizing, snack food manufacturers etc. The marketing centres for these are Maharashtra, Andhra Pradesh, West Bengal and Tamil Nadu.

(6)   *Wafer and Papad* : These are produced in Tamil Nadu for human consumption. The marketing centres for these are Gujarat, Delhi, Maharashtra and Uttar Pradesh.

(7)   *Raw Tubers* : Raw tubers are produced in Kerala, Tamil Nadu, Andhra Pradesh and Assam. They are consumed by the people and is also used for feeding the cattles. Tubers are marketed in Kerala, Tamil Nadu and Andhra Pradesh.

As far as the marketing of starch and sago in the domestic market is concerned, it takes place either by direct sales or through sago serve. The total volume of starch marketed in India during 1998-99 was 90,000 tonnes and in the case of sago, it was 155250 tonnes during the same period as can be observed from Table 3.9.

Table 3.9 : Domestic Sales of Starch and Sago·

Commodity	Year	Through Sago	Direct Sales	Total
Starch	1997-98	75654	18913	94507
	1998-99	72000	18000	90000
Sago	1997-98	105767	43203	148970
	1998-99	112500	42750	155250

Exports

India exports Cassava and its products like raw tubers, flour and meal of sago, starch of manioc and sago and tapioca and substitutes to countries like Kuwait, Netherlands, Saudi Arabia, UAE, Oman, European nations and even to USA. These products are exported through different ports. The Cochin port handles frozen Cassava and it is exported to the Gulf countries. The commodity exported through Kakinada port is Cassava dried chips to Italy, Belgium and other European countries. Sago, Starch and sago papad are exported to New York, Sri Lanka and Australia through Chennai port and the Mumbai port also handles these but the destinations are the Gulf countries, Australia and U.S.A. while through Kolkata port sago is exported to Bangladesh.

The total volume of raw tubers exported during 1997-98 was 49,320 tonnes; in this, a major portion of 34,350 tonnes was exported to UAE alone. In the case of flour and meal of Sago and manioc, the volume of exports during the same period was 160,030 tonnes of which 53,870 tonnes was imported by U.S.A. alone, Malaysia is the main importer of Indian starch of Manioc and Sago. The total volume of exports of these during 1997-98 was 1262,476 tonnes. Apart from these, about 60,250 tonnes of Tapioca and substitutes were exported from India during this period. Oman is the major consumer of these followed by the USA and Kuwait (Table 3.10).

Table 3.10 : Export of Cassava and Its Products from India 1997-98

(in tonnes)

Commodity	Volume
Raw Tubers	49320
Flour and Med of Sago and Manioc	160,030
Starch of Manioc and Sago	1262,476
Topioca	60,205

Source : Foreign Trade Statistics of India.

Problems

(1) Non-availability of qualitative high yielding hybrid Cassava varieties. The quality of the available Cassava hybrids are poor as compared to the local popular cultivars. As a major portion of the Cassava production goes for direct consumption, here the quality of the product is an important aspect.

(2) Fequent attack of diseases especially the Cassava mosaic disease, tuberrot etc.

(3) Lack of effective research towards evolving varieties with shorter duration, drought tolerance, shade tolerance, resistance to diseases and pests, suitable cropping systems for both upland and lowland etc.

(4) Poor resource base of the farmers limits the scope for the adoption of proper technologies.

(5) Poor shelf life of Cassava tubers.

(6) Lack of information on the price behaviour, market demand and marketing channels.

(7) Absence of proper marketing system.

(8) Lack of awareness about the product diversification.

(9) Lack of well defined policy from the side of the Government in India reduced the scope for its improvement. In countries like Thailand and Indonesia, the Government policy supports in this regard.

(10) Lack of organised efforts in tapping the export potential of Cassava products.

Prospects

The use of Cassava as, a human food in the form of value added is increasing in recent years and is becoming a major industrial corp, there is vast scope for area expansion in our country. The advantages from this viz., the rate of dry matter is the hightest among all the crop plants and can be grown under a wide variety of climatic and soil conditions, its rich sources of energy, vitamins, minerals etc. All of these call for an increase in the production of Cassava in India. In this regard, the following steps are useful :

(1) Adopting of high yielding varieties which are disease-free and with high starch content.

(2) Need to identify and supply short duration early bulking varieties.

(3) Researh should be directed towards water, nutrient and drought management to augment productivity.

(4) Need to develop Cassava based value added products like Cassava as a plasti-crop, for binding in fish feed and tissue culture etc.

(5) Need to popularise the use for convenience foods such as chips, wafers and breakfast foods.

(6) Integrated management strategies are needed to overcome the problem of diseases.

(7) As the demand for liquid glucose and dextrose is increasing in food and pharmaceutical industries, the scope for converting Cassava into these is more, hence efforts are needed to make use of these opportunities.

(8) Need for organised efforts to tap the export potential. In this regard, studies are needed on market assessment for export potential exploration, on export demand assessment of Cassava based products and on policy issues for the development of exports. Spot surveys on pre and post-harvesting aspects are needed so as to assess the future demand for Cassava.

Cassava has so far been utilised mainly as fresh food and for the production of starch and sago. Even if there is a change in te food habits of the people pushing Cassava as a secondary food crop in the traditional pockets which has come down drastically, may not dip down further as per the trend observed in the last few years. The country may requre a minimum of 10 million tonnes of Cassava by 2025 as per the present demand and its growth rate and the starch and sago production may be more than 5 lakh tonnes in the coming 25-30 years. As a whole, there is a vast scope for Cassava in our country. In order to promote the use of Cassava and for its own improvement, what it calls in the diversification for alternative uses. Research attempts are needed to develop starch extraction technologies and to modernise the age-old equipments in these factories. Further, there is an urgent need to extend Cassava technologies to non-traditional areas.

Cassava has a number of attributes like an efficient carbohydrate producing crop, tolerant to low soil fertility, has the ability to recover from biotic stress and can be adaptable to multispecies agricultural cropping system and is an alternative crop for small scale farmers with limited resources in marginal areas. By considering all these advantages, there is the need to improve this sector in our country.

SWEET POTATO

Sweet Potato (*Ipomea batata*) is a native of tropical America. Now it is widespread in all tropics and some part of the temperate zone. The Sweet Potato is a twining, trailing perennial vine with adventitious roots that end in swollen tubers. It requires a sandy soil and a warm moist climate. The plants are grown as annuals and propagated vegetatively by using vine cuttings. It is one of the most drought resistant vegetables.

Values and Uses

It's main use is for human consumption. It is eaten usually after boiling and baking. It is also used for canning, dehydration and flour preparation. The tubers of Sweet Potato are good source of starch, glucose, pectin, sugar syrup and industrial alcohol.

It contains 16 percent starch, 4 per cent sugar, that is, 20 per cent alcohol producing material. The starch is 60-70 per cent amylopectin and 30-40 per cent amylose, while the protein constitutent has sufficient amounts of all essential acids required in the diet. The tops are often used as green vegetables and nutritients as it is rich in Vitamins A, B and C. The leaves have 2.35 per cent protein, 4.99 per cent fibre, 1.35 per cent phenols and 0.33 per cent amino acids like Cysteine, Methionine and Lysine. The protein content is as high as 5 per cent on fresh weight basis but is is not easily soluble or extractable calcium, phosphorus and iorn are high in the leaves and they are rich source of ascorbic acid, carotene and vitamin B.

In the world, Sweet Potato is grown in an area os 8937 thousand hectares with a production of 1,19,908 thousand tonnes. Of this, nearly 80 per cent is grown in Asia and China is it's largest producer and it's share is more than 83 per cent in the world production. The other major producers are Uganda, Indonesia, Vietnam, India. Over the years, the production of Sweet Potato has been declining as can be observed from Table 3.11.

Area and Production in India

In India, it is grown mostly in Orissa, Bihar, Uttar Pradesh, Madhya Pradesh and Maharashtra states. Orissa ranks first both in-terms of area and production followed by Uttar Pradesh, Bihar, Maharashtra and Madhya Pradesh. Table 3.12 gives data on state-wise area, production and productivity of sweet potato in major producing states in India.

Table 3.11 : Production of Sweet Potato in Major Producing Countries of the world

(in 000 tonnes)

Country	1993	1994	1995	1996	1997	1998	1999
China	105188	105180	117606	125004	120204	104208	100208
Uganda	1978	2151	2223	1548	1894	2176	2520
Vietnam	2480	2541	1686	1697	1697	1517	1517
Indonesia	2088	1854	2171	2029	1900	1928	1928
Tanzania	260	267	451	420	336	403	403
India	1185	1150	1128	1174	1174	1200	1200
Philippines	693	700	579	595	595	568	568
World Total (including others)	124089	124339	136273	143142	138425	123575	119908

Teble 3.12 : Area, Production and Productivity of Sweet Potato in Major Producing States in India (1996-97)

States	Area ('000 ha)	Production ('000 tonnes)	Productivity (tonnes/ha)
Andhra Pradesh	2.20	13.40	6.09
Assam	9.00	31.70	3.52
Bihar	17.20	156.50	9.10
Haryana	0.10	2.60	26.00
Karnataka	4.00	30.90	7.73
Kerala	2.00	21.00	10.50
Madhya Pradesh	6.60	41.80	6.33
Maharashtra	5.40	76.70	14.20
Meghalaya	5.20	16.90	3.25
Orissa	46.20	326.20	7.06
Tamil Nadu	0.60	12.50	20.83
Uttar Pradesh	27.20	301.60	11.09
Total (including others)	130.30	1061.80	8.15

In Orissa sweet potato is grown in the districts of Koraput, Sundergarh, Sambalpur and Kalahandi, while in Uttar Pradesh Etah, Badaun, Farukkabad and Sultanpur districts as main areas. Bihar has 16 per cent of the area and 18 per cent of the production of Muzzaffarpur, Champaran and Ranchi districts.

The cropping system involving sweet potato vary from region to region. In Orissa, the common sequence are maize-sweet potato follow and Paddy-sweet potato-follow. In Uttar Pradesh it is grown usually as a 'Kharif' crop in sequences to cereals. In drought prone aras of North Bihar plains and the 'Diara' areas, it is favoured.

As a whole, the total area under sweet potato in India has been declining over the years and the same is in the case of production also as can be seen from Table 3.13. However, according to NCA, the area under sweet potato will exceed 2 lakh hectares in India by A.D. 2000, but still it has to be realised. In India sweet potato is generally grown by small farmers on marginal land as a subsistence crop for immediate consumption.

Table 3.13 : Area, Production and Productivity of Sweet Potato in India

Year	Area ('000 ha)	Production ('000 tonnes)	Productivity (tonnes/ha)
1970-71	206.00	1731.60	8.40
1975-76	246.33	1709.95	6.94
1980-81	208.60	1501.80	7.20
1985-86	174.75	1403.60	8.03
1990-91	151.45	1202.63	7.94
1991-92	155.00	1250.00	8.06
1992-93	141.40	1215.50	8.60
1993-94	143.90	1220.60	8.48
1994-95	138.10	111127.80	8.17
1995-96	140.70	1138.10	8.09
1996-97	130.30	1061.80	8.15
1997-98	141.00	1174.00	8.32
1998-99	144.00	1200.00	8.33
1999-2000	144.00	1200.00	8.33

Source : Upto 1997 D of E&S for 1998-2000, FAO.

Varieties

Several varieties of sweet potato are available and cultivated throughout the world. The varieties are grouped according to their colour, which varies from white, golden and orange to red. People in various parts of India seem to have a traditional preference for white or red colour of the skin of the tuber. In Karnataka, it is the white variety which is preferred while in other parts of India, both red and white are popular.

The most important varieties grown in India are : (1) Pusa Suffaid which is high-yielding and produces medium-sized tubers; (2) Pusa Fal, a Japanese variety, its skin colour is red but the cooked flesh is white; (3) Pusa Sandheri is a selection from the U.S.A. The flesh colour of this variety is light orange and contains higher carotene than the other varieties. Apart from these, several other varieties are grown in different states. In Bihar SP-3, SP-9 are popular, in Karnataka viz., Hosur Red and Hosur Green, in Tamil Nadu V6, V8, V12 etc. are popular while in West Bengal Ranger and B-4 306 are grown.

Harvesting

Harvesting of sweet potato takes place when the leaves turn pale and latter turn slightly yellow. After harvesting, they are cleaned and graded for the market.

Marketing

As there is no organised marketing system for sweet potato, most of the growers sell this to the commission agents or contractors where the scope for expolitation is large. The growers sell this to these intermediaries or contractors since the product is bulky and requires storage facilities, all of these are beyond the reach of small farmers. Hence the price received by the producers are below 50 per cent in the consumers price. The major utilization pattern of sweet potato is as vegetable and subsidiary food in the case of weaker sections and tribals.

Problems

(1) Non-availability of high yielding short duration quality hybrid planting materials.

(2) Sweet potato weevil is a big menace in its cultivation.

(3) Non-availability of proper marketing facilities.

(4) Lack of storage facilities.

(5) Lack of stability in tuber production.

(6) Low return per unit area, shift in cultivation towards other crops.

(7) Lack of awareness about nutritional value and change in food habit.

(8) Limited product diversification and lack of processing industries.

(9) Lack of government policies and programmes for an improvement and development of sweet potato cultivation.

Prospects

As sweet potato is used in food and can be used in feed and in the preparation of value added food products, there is the need to improve the status of this crop. As a mixed crop, it can increase the earnings of the small and marginal farmers and has been playing an important role in improving the living standards of the poor and weaker sections of the society, there is an urgent need to have a need based plan for its development. In this regard, the following measures will be useful. They are :

(1) There is the need to produce disease free short duration high yielding planting material.

(2) Studies are needed to assess the causes of non-tuberization, uneconomical yield performance as well as integrated nutrient management.

(3) Need to identify the most compatible and profitable cropping system involving sweet potato in major sweet potato growing states.

(4) Concern on the public health hazards of synthetic colouring agents is increasing day by day which has grdually led to an increased efforts to extract natural pigments, which are safe for use in food products. In this regard, sweet potato is a promising crop with lot of variability in pigment content in germplasm collections. This can be favourably put to use in selecting high pigment cultivars and also developing cultivars with enhanced pigment levels.

(5) Need to identify and develop the other uses of sweet potato especially in the preparation of value added foods. Efforts are needed for extraction of anthocyanins for food industries. Apart from these, its uses in pharmaceuticals has to be identified.

(6) Studies on market assessment for export potential exploitation is required.

(7) Development and extension of technology.

(8) Governmental support to improve the existing situation and to develop this sector through proper planned programmes.

(9) Need to improve the marketing system along with proper storage and other facilities.

YAMS

Yams belong to the genus Dioscorea. It is very difficult to distinguish Yams since there are a great many species in the tropics and subtropics of all countries. The most commonly cultivated species is *D. alata*. Yams are all climbing vines with large storage roots. They require a deep soil, but are quite drought-resistant.

Values and Uses

Yams are a low energy, high protein and vitamins A and C food. They are the chief food of millions of people in the West Indies, South America and the Asian tropics. They are baked or boiled or ground into flour. In southern U.S., they are used to feed hogs and other livestock. In Western Africa, Yams enjoy certain social value. In India, Yams are moslty used as vegetable.

Area and Production of Yams in the World

Yams are grown mostly in the African region. The leading producers are Nigeria, Cote Divoire, Ghana Benin, Ethiopia PDR, Sudan and Tago. Nigeria ranks first both in terms of area and production of Yams in the World and its share is about 69 per cent of the total in both area and production. The total area under Yams in the world in 1999 was 3805 thousand hectares with a production of 36033 thousand M.T. As a whole, the African region consists of 96 per cent of the area and it contributes the same percentage towards production in the world as can be observed from the Table 3.14.

In India, some of species are grown in eastern Uttar Pradesh, Bihar, and South Indian states. They are grown mostly under mixed cropping system in Eastern India, Kerala and in some tribal pockets. By intercropping, farmers reduce risks associated with disease, pest and price fluctuations, stabilise the flow of food and income throughout the year and intensify land use in our country. As Yams are generally inter-cropped, the production figures are either misleading or not available.

Table 3.14 : Area and Production of Yams in Different Regions of the World

(Area in '000 ha, Production in '000 M.T.)

Region	1997		1998		1999	
	Area	Production	Area	Production	Area	Production
Africa	3144	28797	3647	34240	3662	34430
N.C. America	74	622	60	471	60	471
South America	49	487	49	487	49	487
Asia	16	234	16	234	16	234
Oceania	18	272	18	271	18	272
World Total (including others)	3301	30451	3790	35843	3805	36033

Source : FAO.

Problems

The major problems in the production of Yams are their longer duration and requirement of large quantity of tubers as seed materials. Lack of standardized system of cropping reduces the scope for increasing the productivity of Yams in India. Along with these storing of Yams is also a problem, which accounts for more than 50 per cent of cost of production. Problem of pests and diseases both in the field and storage conditions reduce the scope for improving this crop in our country.

Prospects

As Yams are used as vegetables and have medicinal properties and their extract is used for treatment of arthritis, there is an urgent demand to develop this in India. In this regard, the following steps will be useful :

(1) Need to undertake steps to standardize agro-techniques of Yams in relation to the existing cropping system as well as sole crop of Yams.

(2) Steps should be taken on production, evaluation, multiplicaton and distribution of quality planting materials of Yams.

(3) So as to reduce the cost of cultivation, isolation of superior dwarf varieties is needed.

(4) Need for appropriate strategies to overcome the problems of pests and diseases.

(5) Need for farmers' participatory research on pre and post-harvest technologies.

(6) Need to conduct frequent surveys on area, production and post-harvesting aspects.

ROOT CROPS

Carrot, radish, beet and trunip are the important commercial crops grown under this group. Apart from these, some minor crops like skirret, parsnip, rutabaga etc., are also included in this category. All of these thrive well in a cool season. However, carrot, radish and trunip are grown in warm season too.

These are rich in minerals and vitamins and are consumed by the people all over India. Even though the state-wise as well as national figures for area, production and yield for these crops are not available, still for an understanding, a breif discussion is made here on Carrot and Radish.

CARROT

The carrot (*Daucus carota*) has been cultivated for over 2000 years. It was known to the Greeks and Romans and reached Europe early in Christian era. It was a favourite vegetable in England during the time of Queen Elizabeth and was brought to Virginia and England. The Indians carried it over the rest of America. Carrot probably originated in Central Asia, now it is found all over the world.

Values and Uses

Carrot leaves are nutritive rich in protein, minerals and vitamins. It contains 86 g moisture, 0.9 g protein, 1.2 g fibre, Vitamin A and Vitamin C. Carrot is used for human consumption as well as for forage. They are eaten raw or cooked and are often used for flavouring soups and stews. It is also used in the preparation of pickles and sweet-meat. The yellow colouring matter, carotin is sometimes exracted and used for colouring butter.

Area and Production

The total area under carrot in the world at present is 795 thousand hectares and the production is 18,303 thousand M.T. China ranks first both in terms of area and production of carrots followed by USA, Poland, Nigeria, Japan and India in terms of area while in terms of production next to China are USA, Poland Japan, France, U.K. and India. The total area under carrots in China is 203 thousand hectares with a production of 4478 thousand M.T. in terms of area as well as production in the world, it has been more or less stable since 1997 as can be seen from Table 3.15.

Table 3.15 : Area and Production of Carrots in Major Producing Countries in the World

(Area in '000ha, Productn in '000 M.T.)

Country	1997		1998		1999	
	Area	Production	Area	Production	Area	Production
China	203	4478	203	4478	203	4478
USA	54	2266	55	2201	55	2201
Poland	30	799	33	992	30	925
Nigeria	25	228	26	229	26	234
Japan	24	715	24	720	24	720
India	24	340	24	340	24	340
Indonesia	17	227	15	260	15	260
U.K.	16	625	16	618	16	618
France	16	652	16	672	16	670
Algeria	10	125	10	141	10	140
World Total (including others)	805	18415	799	18356	795	18303

Source : FAO.

In India, carrot is grown in almost all areas. The major growing areas are Panjab, Uttar Pradesh and Madhya Pradesh. The total area under carrot in India is about 24 thousand hectares an the production is 340 thousand M.T. India stands fifth in terms of area under carrot in the world while in terms of production, it ranks seventh.

Varieties

The important varieties gown in India are Pusa Kesar, Pusa Meghali, Indian old etc., As a whole, the varieties are classified into two groups viz., temperate types and tropical types. The tropical types usually give higher yield and it varies in between 20 to 30 thousand kg per hectare.

Future

As far as carrot and its cultivation in India is concerned, the work carried for its improvements is minimum. There is the need to undertake a detailed survey about the present situation and future prospects. In this regard, the organisations have to take the required steps. There is also the need to improve the tropical varieties in our country.

RADISH

Radish (*Raphanus sativus*) is grown all over the world and are highly esteemed because of their pungent flavour. It is an important root vegetable grown in India for its fleshy roots and gree foliage. It is grown throughtout India. The leading producers are Uttar Pradesh, Punjab, Maharashtra and Gujarat.

Value and Uses

Radish is very rich in minerals and vitamins A and C. It is eaten raw or as a salad or cooked. The roots and leaves of this vegetable are not only nutritious but also abundant and relatively low priced in the market compared with other root vegetables, hence it is consumed by all sections of people. It is very refreshing when eaten fresh-full of ascorbic acid and a variety of mineral salts. The taproot vegetable is effective in curing liver, gall-bladder and urinary disorders, piles and gastrodynia. The seeds of this yield hava non-drying fatty oil suitable for soap making.

Varieties

The varieties of radish can be classified into two viz., European type or temperate type and the Asiatic or tropical and sub-tropical types. The first on is small in size, mild and are used for *salad* purposes. The important varieties are white Icicle, Pusa Himan, Pusa Desi, Pusa Chetki etc.

Future

The development work carried out for the improvement of radish in India is insufficient, hence efforts are needed on these lines. As it is having nutritional as well as medicinal values, there is an urgent need to popularise this in our country. In this regard, there is the need to educate the farmers on various cultivation practices and post-harvest activities.

ONION

Onion belongs to the family of Amaryllidaceae, genus Allium and species *Cepa*. The genus *Allium* contains about 300 species, biennials and perennials, mostly bulbous. It is the chief food plant in which the food is stored in a bulb. It is very old, its use going back over 4000 years, beyond the beginnings of authentic history. It is unknown in the wild state. It was probably a native of South Asia or the Mediterranean region. This crop was know for its various uses in Egypt since 3200 B.C. and in India since 600 B.C. In Egypt, it was worshipped before the Christian era. Onion are cultivated over large areas in temperate and even in tropical climates. It prefers cool moist regions with a sandy soil. Onions are grown from seeds or sets, small bulblets that are produced instead of flowers.

Value and Uses

A big onion contains 86.8 per cent moisture, 1.2 per cent protein, 0.1 per cent fat, 11.6 per cent carbohydrates, 0.18 per cent phosphorous, 0.05 iron etc. It is used both for cooking and as a condiment for flavouring or for pickling. Mild onions are used for cooking or as *salad*. Pungent varieties are used as condiment for flavouring a number of foods. Pearl onions or small onions are used in pickels including vinegar pickles. Onions are baked, boiled, fried and used in fresh or dehydrated or powder form in soups, sauces etc., they are also eaten raw as a *salad*. Onion is said to possesses stimulant, diuretic and expectorant properties and is considered useful in flatulence and dysentery.

Area and Production in the World

The notable onion growing coutnries in the world are China, India, the U.S.A., Turkey, Japan, Iran. In China, it is cultivated in an area of 451 thousand hectares and its production is about 10040 thousand M.T. India ranks second both in terms of area and production while in U.S.A., the area under onion is below than that of Turkey, Russian Fed, Pakistan and Iran still in terms of production stands thrid. Table 3.16 gives data for area and production of onions in the major producing countries in the world.

Table 3.16 : Area and Production of Onions in Major Producing Countries in the World
(Areas in '000 ha, Production in '000 M.T.,

Country	1997		1998		1999	
	Area	Production	Area	Production	Area	Production
China	451	10040	451	10040	451	10040
India	415	4429	415	4429	415	4429
U.S.A.	67	3119	67	2995	67	2995
Turkey	105	2100	105	2300	105	2300
Japan	27	1256	27	1240	27	1240
Iran	46	1157	48	1210	48	1210
Pakistan	81	1131	81	1077	81	1077
Russian Fed	94	1077	103	1054	103	1054
Spain	26	959	24	981	24	985
Korean rep	11	740	11	872	11	936
World Total (including others)	2328	39308	2334	33912	2332	40023

Source : FAO.

As a whole, the total area under onion has been remaining more or less same while the production is increasing over the years. The total area under onion in the world in 1999 was 2332 thousand hectares and the production was 40023 thousand M.T. of the total area under onion in the world, the share of Asian region is 60 per cent and in terms of production also it contributes the same amount. As far as the share of China is concerned, it is having 19 per cent of area under this crop in the world and its share in world's production is 25 per cent while India's share in the area is around 18 per cent and in production it is 11 per cent.

Area, Production and Productivity in India

Onion is one of the most important commercial vegetable crops grown in India. It is cultivated in Maharashtra, Karnataka, Orissa, Uttar Pradesh, Tamil Nadu and Andhra Pradesh on a large scale while in other states also the cultivation of it is undertaken, The total area under onion in Maharashtra is around 65 thousand hectares and the production is around 573 thousand tonnes. In

Karnataka, it is cultivated in an area of 61 thousand hectares and its production is around 360 thousand tonnes. In terms of productivity, it is the maximum of 29617 kg/ha in Gujarat followed by Punjab, Haryana, Andhra Pradesh, Madhya Pradesh as can be seen form Table 3.17.

In Maharashtra, Nasik district alone contributes a major portion of onions in the state's production; in Karnataka, Bangalore, Hubli, Devanagere districts are the main contributors; in Andhra Pradesh, Cuddapah, Ananthapura district's share is more. The districts of Alwar, Jodhpur, Ajmer, Jaipur are the major contributors in Rajasthan while in West Bengal Hooghly district is the main production area.

As a whole, the total area under onion at present is about 415 thousand hectares in our country with a production of 4429 thousand tonnes. In terms of area as well as production and productivity of onions, it has been fluctuating over these years as can be seen form Table 3.18. The area under onion in 1996-97 was estimated at 4.04 lakh hectares as against 3.96 lakh hectares estimated in 1995-96. The area increased 2.02 percent during 1996-97 over 1995-96, however, in terms of area, it came down during 1997-98 to 3.39 lakh hectares which shows a negative growth, while in terms of production also it came downwards and similar is the case with productivity too.

Varieties

The varieties of onion are classified into three viz., (1) *Desi* varieties like Ratna White, Patna Red, Nasik Red, Red Globe, White Globe etc., (2) improved varieties like Pusa white round etc., (3) Exotic varieties like silver skin, Red Italian, Sweet Spanish, White Portugal, Texas Early Garnno (Yellow type) Australian Borwn etc., As a whole, two types are normally grown, they are Red onion varieties and White onion varieties.

Important Red Onion Varieties

(1) *Pusa Ratnar* : It is a bronze deep red colour, obvate to flat globular, less pungent, neck dropping. The quality is good for storage purposes.

(2) *Pusa Red* : It is a very popular short-to-intermediate day length variety, its bulbs are medium in size, bronze in colour, flat to globular shaped and less pungent.

Table 3.17 : State-wise Area, Production and Yield of Onion from 1992-93 to 1997-98

(Area '000 ha, Production '000 tonnes, Yield – kg/ha)

State	1994-95			1995-96			1996-97			1997-98		
	Area	Prodn.	Yield	Area	Prodn.	Yield	Area	Prodn.	Yield	Area	Prodn.	Yield
Andhra Pradesh	20.8	314.1	1510	24.00	373.6	15567	27.2	369.0	13566	21.3	328.3	15413
Bihar	20.0	154.0	7700	17.5	132.5	7846	19.1	145.2	7602	18.3	140.7	7688
Gujarat	20.3	561.5	27660	19.00	444.00	23368	16.5	433.0	26242	20.9	619.0	29617
Haryana	2.4	41.6	17333	6.3	106.3	16873	4.5	607	13488	1.7	26.9	15823
Himachal Pradesh	0.7	1.5	2143	0.8	4.3	5375	0.8	4.6	5750	0.8	2.9	3525
Karnataka	66.7	489.1	7333	78.8	440.3	5588	90.5	508.7	5620	60.4	359.9	5959
Madhya Pradesh	17.7	181.2	10237	21.1	235.4	11156	23.6	308.9	13088	20.2	261.9	12965
Maharashtra	101.6	1210.0	11909	92.60	1120.50	12100	95.5	1189.1	12451	64.8	572.2	8830
Orissa	43.8	320.0	7306	50.00	380.00	7600	37.5	295.4	7877	47.0	165.0	3511
Punjab	2.5	50.3	19346	1.6	43.30	20125	3.1	61.0	19677	2.5	48.0	19200
Rajasthan	16.6	83.2	5012	21.70	161.70	7450	24.4	206.2	8450	19.6	112.7	5750
Tamil Nadu	34.4	302.5	8794	21.40	235.50	11005	24.9	210.2	8442	26.6	235.0	8835
Tripura	0.2	0.3	1500	0.1	0.2	2000	0.1	0.2	2000	0.1	0.2	2000
Uttar Pradesh	29.5	332.5	11271	33.5	396.1	11824	29.7	386.1	13000	26.5	252.1	9513
All India (including others)	384.4	4057.9	10556	395.5	4080.0	10316	404	4181	10349	338.5	3142.8	9284

Source : Economics and Statistics : Adviser, New Delhi.

Table 3.18 : Area, Production and Productivity of Onions in India

Year	Area	Production	Productivity
1980-81	2.51	25.05	9,961
1985-86	2.80	28.60	10.202
1990-91	3.02	32.25	10,686
1991-92	3.23	35.84	11,088
1992-93	3.23	34.90	11,168
1993-94	3.68	40.06	10,902
1994-95	3.79	40.36	10,661
1995-96	3.96	40.80	10,316
1996-97	4.10	40.39	10,834
1997-98	3.94	36.86	9,355
1998-99	4.15	44.56	10,733
1999-2000	4.15	44.29	10,672

Source : Upto 1998-99 Ministry of Agriculture, for 1999-2000 FAO.

(3) *Akra Niketan :* The bulbs of this can be stored for 5 months under room temperature. It can be grown both in rabi and kharif. It has attractive colour and its neck is thin.

(4) *Arka Kalyan :* It is moderately resistant to the purple blotch disease with bulbs globose shaped with deep pink coloured outer scales. It is a suitable variety for the kharif season.

(5) *Agrifond Dark Red :* It is identified for the plains of Sutlej-Ganga region. The bulbs of this are red, globular in shape with tight skin, moderately pungent. It cen be stored for a longer period compared with other varieties. Among the Kharif grown varieties, it has the best strong capability.

(6) *Punjab Selection :* Its bulbs are red, globular in shape and bulbs are firm with good keeping quality and is suitable for dehydration.

Important White Onion Varieties

(1) *N-257-S-1 :* The bulbs are globe shaped, white. It has good keeping quality and a high yielding one. It is suitable for rabi crop in onion growing areas of Maharashtra.

(2) *Pusa White Round* : This variety is suitable for dehydration. The bulbs are white and round shaped.

(3) *Pusa White Flat* : The bulbs are attractive and flattish and medium to large in size. It is suitable for dehydration and keeps healthy in storage.

(4) *Udaipur 101* : It is a recommended variety for Rajasthan and the adjoining states. Bulbs are deep red, flattish, globular and sweet with less pungencey. It is good for *salad.*

Apart from these, some of the other high yielding varieties like 548, Udaipur 103, Hissar 2, Agrifond Light red, Kalyanpur Red Round, Arka Pragati etc., are also grown in India. The hybrids like BSS214, BSS 100, exotic varieties like V12, hybrids such as V7, VII, VI6, Deco 551 are also cultivated in our country. In Karnataka, the Bangalore rose, Podisu, Nattu and Krishnapuram variety onions are specially grown for exports. They are not consumed in India but command a premium in the markets of Singapore and Malaysia.

Harvesting

The duration of local varieties is usually 4 months after transplanting while in the case of introduced varieties, it is 15 days more than the local varieties. After harvesting curing, washing and further processing takes place.

Marketing

The marketing technology of onion has not received as much attention as the production technology in our country. Unless the marketing efficiency improves, no incentives to increase production will attract the cultivators of onion. Better returns, stable prices and attractive opportunities to trade will induce the growers to produce more and market in an increasing proportion of their produce. It is common to observe that the growers of onion do not get remunerative prices, while the consumers have to pay higher prices for the same. Absence of proper marketing and storage has actually made for an excess loss or waste as can be observed in the streets of Nasik as well as in other areas where onion is grown.

Even though onion is covered under market legislation, in most of the growing areas, there appears to be the prevalance of two major channels in its marketing. They are:

(1) Producer-seller-commission agent-cum-wholesaler - Retailer-Consumer.

(2) Producer-seller-village merchant commission agent-cum wholesaler Retailer-consumer.

Between these two channels, sale through commission agents-cum-wholesalers is most popular. In most of the cases, in its marketing, the producers' share is around 65 per cent while the remaining percentage goes to the intermediaries. The important marketing centres which supply onion to the domestic markets are Bangalore, Calcutta, Mumbai, Mangalore, Hyderbad, New Delhi, Bellary etc.

Price Trend

During the year 1997-98, the prices of onion ruled low till September 1997, thereafter the prices showed rise and the trend continued till January when the index number of wholesale prices of onion increased to 883.9. From February 1998, the prices again started declining and the index number declined to 349.6 in April 1998, but the prices again shot up in May, 1998 and the trend continued till the end of agriculture year when the index number of wholesale prices of onion reached to 471.3 as can be observed from Table- 3.19. It is interesting to note that in the hisotry of onion

Table 3.19 : Index Number of Wholesale Prices of Onion
(Monthly average base 1981-82=100)

Month	1996-97	1997-98	Percentage variation 1997-98 / 1996-97
July	343.9	277.2	−19.4
August	299.3	270.9	−9.5
September	278.7	235.6	−15.5
October	311.1	325.8	+4.7
November	342.2	361.4	+5.6
December	348.6	570.6	+63.7
January	333.7	883.9	+164.8
February	299.7	776.9	+48.5
April	330.7	349.6	+5.7
May	308.4	409.9	+32.9
June	257.8	471.3	+82.8

market, a record price was observed on February 3, 1998, when the price of top quality onion in the Lasalgaon market in Nasik in Maharashtra touched as high as Rs. 1,851 per quintal and for the poor quality onions, it was Rs. 1,451-1,500 per quintal.

The retail price of onion in Delhi market went up from Rs. 6-8 in October to Rs. 12 in November, Rs. 18-20 in December 1997 and the maximum of Rs. 25 per kg. in January 1998. In 1998, the Union Government asked NAFFD to procure onions and the NDDB was also asked to take market intervention operations. The Delhi government even purchased onions at Rs. 40 per kg. and distributed it to the citizens at Rs. 10 per kg. As a whole, the price of onions in major consuming centres in India went up to Rs. 40-60 per kg during Oct-Nov. 1998. However, the price of onions came down in January 1999 to Rs. 250-500 per quintal in the major marketing centres.

Market Intervention Scheme

To protect the interest of the farmers from distress sale of their products, the Government has brought Market Intervention Scheme (MIS) for onion. The National Agricultural Cooperative Marketing Federation of India Ltd., (NAFFD) is a Centrally designated nodal agency for market intervention operations. The Profit/Loss incurred, if any, in these operations are shared on 50 : 50 basis both by the Central Government and the concerned State Governments.

Exports

Among the horticultural products, onion is the single largest commodity exported from India. It accounts for about 90 per cent of the exports of vegetables from India in terms of value. India exports fresh and chilled onions to UAE, bangladesh, Singapore, Malaysia, Sri Lanka, Russia, Pakistan, Mauritius, Maldives and Kuwait. The total volume of exports during the period 1997-98 was 3,33,349 M.T. valued at Rs. 202.46 crores. UAE is the major importer of our onions followed by Malaysia, Sri Lanka and Bangladesh as can be seen from Table 3.20. The trend in our exports of onions since 1980-81 up to 1997-98 as shown in Table 3.21, indicates that it has been moving in the positive direction. However, frequent fluctuations are common in its exports since the internal production also shows the same type of trend over these years.

Table 3.20 : Country-wise Exports of Fresh or Chilled Onions
(Volume in M.T. and value Rs. Crores)

Country	1996-97		1997-98	
	Volume	Value	Volume	Value
Baharin	3,621	1.85	1,633	1.29
Bangladesh	61,635	36.29	50,035	25.97
Kuwait	7,374	3.84	5,067	2.70
Malaysia	99,180	6?.71	78,377	50.96
Mauritius	5,919	3.90	5,096	2.93
Saudi Arabia	16,622	8,69	13,114	9.23
Singapore	36,210	30.48	32,441	30.21
Sri Lanka	81,509	50.53	57,208	·28.87
UAE	1,08,201	61.31	85,532	46.63
Total	4,27,012	265.21	3,33,349	202.46

Source : DGCIS.

Table 3.21 : Export of Onions from India
(Volume in lakh M.T. Value in Rs. Crores)

Year	Volume	Value
1980-81	1.94	5.34
1985-86	1.57	29.21
1990-91	2.79	84.55
1991-92	4.16	116.00
1992-93	2.72	183.23
1993-94	3.57	119.59
1994-95	4.97	182.68
1995-96	4.01	204.62
1996-97	5.13	336.56
1997-98	4.41	293.76

Source : DGCIS.

As far as the preference for onions in the international market is concerned, in Europe and Japan the yellow/brown, less pungent variety of onions is preferred while in the Gulf nations and in South East Asian countries, the light red more pungent variety is preferred. The small pickling quality onions are preferred in Singapore, Malaysia and Sri Lanka. The special varieties of onions, grown in Karnataka, that of small size 25-35 mm. in diameter like Bangalore rose, Podiser, Nattu and Potlore are having an excess demand in the international market especially in Malaysia and Singapore and are sold at a premium price. For example the Bangalore Rose fetches double the popular varieties in terms of value. The major terminal marketing like Nasik, Chennai, Pune, Nagapattanam, Calcutta, Mumbai and Bangalore share the maximum exports of onion from India.

For the export of big onions from India, the Director General of Foreign Trade (DGFT) authorised three canalising agencies namely National Agricultural Cooperative Marketing Federation of India (NAFED), Maharashtra State Government Agency and the Gujarat Government Agency. All these designated canalising agencies export the prescribed volume onions according to the instructions. Recently, because of the onion crisis that of Sept-Nov., 1998, a ban was put on the exports, however, a partial lifting was observed during 1999.

Problems

(1) The productivity of onion in India is much lower when compared with other growing countries of the world. In India, it is 10 tonnes per hectare, while it is above 44 tonnes per hectare in Japan and USA while in China it is above 22 tonnes per hectare. As a whole, our productivity is below the level of world average productivity which is above 17 tonnes per hectare.

(2) Diseases like soil borne diseases, foliar diseases, leaf blight, Twister and diseases of bulbs. Apart from these onion thrips, a common and serious pest in onion production.

(3) Inadequate Irrigation facilities.

(4) Non-availability of high-yielding suitable for different climatic conditions and resistant varieties.

(5) Lack of proper organised marketing system. Even though onions are covered under the market regulations or legislations, still the role of the middlemen is maximum in its marketing. As far as the sales in the wholesale marekts are concerned, the growers of onion are facing several problems in these centres like congestion, lack of physical facilities for display, storing and weighing.

(6) Non-availability of proper storage facilities is mainly responsible for distress sales and wastages. Most of the available storage facilities within the production areas

are traditional in nature which leads to huge losses. Proper storage extends the period of availability of fresh onions by arresting the metabolic breakdown and deterioration caused by micro organism. The need here is to achieve proper storage by controlling the temperature, relative humidity and atmospheric concentration of certain gases, as also by chemical treatment and eradication.

(7)   Lack of proper steps to expand the exports.

Prospects

Considering the present population growth rate and future expectations, the current level of production of onions is insufficient to meet the future demand. Again the present export of onion is almost negligible and the scope for expansion is always there. Along with these, the processing industries are emerging at a faster rate and would demand more raw materials in the future and the seed industry will also consume sizeable proportion of bulbs produced in the future. Taking all these aspects and the estimated demand for onions in India, it may cross 80 lakh tonnes which includes the post harvest losses also, and the estimated exports by the same period will be at least 20 lakh tonnes. All of these give an indication to expand the production and productivity of onions with necessary pre- and post-harvest technologies. So as to have this, the following aspects have to be seriously taken into consideration. They are :

(1)   Increasing the productivity through genetic manipulations and making sufficient quantity of seed of released varieties. Further increasing the productivity of improved varieties/hybrids through better agronomic manipulations.

(2)   There is the need to prevent loss through genetic manipulations, cultural management and improvement in storage and packaging conditions.

(3)   Value addition through processing of the bulbs in the form of dehydrated flakes, dehydrated powder, paste etc.

(4)   Need to develop export oriented variations.

(5) Need to extend the cultivation of onion in the non-traditional areas.

(6) Need to promote the export of onion to European countries. Along with these, care should be taken on quality and price aspects.

(7) Need to provide incentives and proper infrastructural facilities to the existing processing industries. Again, there is also the need to expand these industries.

(8) Steps are needed to promote the export of onion seed.

(9) Need to improve the marketing system. Proper facilities should be made available in wholesale marketing centres. The required number of storage and godown facilities have to be provided to the farmers so as to minimize the losses and to get a remunerative price for onion.

As a whole, what we need is to grow more quality onions and take greater care in post-harvest activities.

GARLIC

The garlic (*Allium sativum*) is a perennial plant with narrow flat leaves and several egg-shaped bulbs, known as cloves, enclosed in a white skin. The inflorescences produce both seeds and bulblets. The history of garlic, like that of onion, dates back to time immemorial. Central Asia is the prime origin centre followed by the Mediterranean area. It has been recognised all over the world as a valuable condiment for foods, and a popular remedy or medicine for various ailments and physiological disorders. It grows under much the same conditions as the onion, except that it favours a richer soil and a higher elevation. Garlic is a cool season crop and it succeeds best in mild season without extreme heat and cold.

Values and Uses

Garlic is quite rich in vitamins 'A', 'B', and 'C'. In the analysis of garlic, it is found to have six and a half per cent proteins, one and a half per cent of iron contents, one per cent each of minerals, fats, lime content and phosphorus and about 30 per cent of carbohydrates, besides other natural elements.

As a condiment, garlic is used practically all over the world for flavouring dishes. In the U.S.A., nearly half of the entire production of fresh garlic is dehydrated. It is used in *salad* dressing, tomato products and in several meat preparations. Raw garlic is used in the manufacturing of garlic powder, garlic salt, garlic vinegar, garliced potato chips, garlic tread etc. In recent years, there has been a growing demand for garlic from the food industries in India.

The Unani and Ayurvedic systems say that garlic is carminative and is a gastric stimulant, and thus aids in digestion and absorption of food. It sets right the wear and tear of the body and invigorates the dead cells of the body. It cures the body of the consumptive ailments like TB, Asthma and the similar diseases. Its powerful odour provides a security cover to our body. In modern allopathy, it is used in a number of patented medicines. The anti-bacterial action of garlic had been noticed from early days, its healing capacity and effectiveness against cholera having been recorded as early as in the eighteenth century.

Its oil is an effective insecticide and is a valubale flavouring agent. Its juice is used for various ailments of the stomach, as a ruberfacient in skin diseases and as ear drops.

Areas and Production in the World

Garlic is mainly grown in China, India, U.S.A., Russian Federation and Spain. The total area under this during 1999 was 889 thousand hectares with a production of 8776 thousand M.T. In terms of area and production, the Asian region stands first and its share is 77 per cent and 83 per cent respectively. China ranks first both in terms of area and the production of Garlic in the World followed by India as can be observed from Table 3.22.

Table 3.22 : Area and Production of Garlic in Major Producing Countries in the World

Country	1998		1999	
	Area	Production	Area	Production
China	424	5690	424	5690
India	113	452	113	452
USA	14	224	14	224
Russia Fed	25	161	25	161
Spain	24	160	25	170

Source : FAO.

Area, Production and Productivity of Garlic in India

Garlic is cultivated practically throughout India. The total area under this crop at present is 113 thousand hectares and the production is about 452 thousand M.T., while the productivity is around 4000 kg/ha. The major producers of Garlic in India are Madhya Pradesh, Gujarat, Orissa, Rajasthan, Uttar Pradesh and Maharashtra. Tabe 3.23 gives data on state-wise area, production and productivity of garlic in India for the period 1997-98 and 1998-99.

Table 3.23 : State-wise Area, Production and Productivity of Garlic in India 1998-99

State	Area (in '000 ha)	Production (in '000 tonnes)	Productivity (kg / ha)
Andhra Pradesh	0.7	1.1	1571
Bihar	2.9	4.2	1448
Gujarat	20.2	134.2	6644
Haryana	0.6	4.4	7333
J&K	0.5	0.4	800
Karnataka	4.1	3.2	780
Madhya Pradesh	39.9	186.8	4682
Maharashtra	6.5	43.6	6708
Orissa	17.5	55.1	3149
Punjab	0.9	10.0	11111
Rajasthan	11.4	33.5	2939
Tamil Nadu	2.0	10.6	5300
Uttar Pradesh	7.2	30.6	4250
All India	114.4	517.7	4525

Source : D of E&S.

The trend in area, production and productivity since 1975-76 up to 1999-2000 shows that it has been increasing remarkably as can be observed from Table 3.24.

Varieties

There is no distinct variety of garlic. However, the available varieties can be grouped into two viz., local varieties like Kanwri, Fawri and Rajalla Gaddi and the Acclimatise varieties like early white or Medican varieties like Ceiole, Tatule and Italian. The speciality of Rajasthan garlic is that the flakes are big and the flavour is good. They are the 'medium' from Madhya Pradesh, 'VIP' quality, 'Bom' quality and the 'Super Bom' quality. The super bom fetches the highest price while the medium the lowest.

Table 3.24 : Area and Production of Garlic in India

Year	Area in Hectares	Production in tonnes
1975-76	39800	140900
1980-81	59300	21600
1985-86	57700	189600
1989-90	83500	327000
1992-93	85500	355800
1993-94	76200	306000
1994-95	98900	403200
1995-96	114800	490000
1996-97	94300	437900
1997-98	98500	464000
1998-99	11440	517700

Source : D of E&S, New Delhi.

Harvesting and Marketing

It is harvested when the top turns yellow to brownish and shows, the signs of drying up. The prevailing marketing system for garlic is not a perfect one. The involvement of intermediaries is common as that of onion. The main marketing centres are located in Mumbai, Gujarat, Bangalore etc., Bangalore is the main exporting centres where both buying and selling of it takes place.

Exports

India exports garlic, dehydrated garlic flakes, garlic powder and paste to South-East Asian countries in larger volume, while for the U.S.A., Canada, EEC's, it is in smaller quantum. During 1998-99, India has exported 4,068 tonnes of garlic volued at Rs. 7.41 crores as against 3,975 tonnes valued at Rs. 7.98 crores of the previous year. The export in 1998-99 includes 1116 tonnes of garlic flakes and 268 tonnes of de-hydrated garlic powder along with 2,685 tonnes of raw garlic from India. Garlic has been mainly exported to Bangladesh, Philippines, Japan, Pakistan and U.K. Table 3.25 gives data for the export of garlic from India since 1960-61 upto 1998-99.

Problems

(1) Non-availability of high-yielding improved varieties and sufficient quality of seed.

(2) Harvesting and post-harvesting problems like the prevalance of intermediaries in its marketing, non-availability of proper transport facilities and absence of storage facilities. Because of these problems, it is noted that about 20-25 per cent of the fresh crop is wasted due to respiration, transpiration and micro-biological spilage.

(3) Under utilization of undersized or cull but healthy bulbs.

(4) Problem of pests and diseases.

(5) Non-availability of proper training and education to the farmers with regard to pre- and post-harvesting aspects.

(6) Failure to find out alternative uses of garlic in a value added manner.

Table 3.25 : Export of Garlic from India

Year	Quantity	Value (Rs. '000)
1960-61	2030	1123
1965-66	795	713
1970-71	1634	2782
1975-76	931	2549
1980-81	7398	22824
1985-86	2619	13568
1990-91	4646	32772
1992-93	7442	71018
1993-94	2844	35489
1994-95	633	12287
1995-96	3936	49125
1996-97	4889	79774
1997-98	3975	79706
1998-99	4068	74108

Source : DG of CI&S, Calcutta.

India ocne a market leader, is losing out in the overseas markets to China, the current market leader. So, efforts are needed to tackle the above said problems and to supply qualitative garlic into the international market. In this regard, the following steps are useful :

(1) There is the need to identify new varieties with large bulbs of white colour, uniform shape and size fold and compact cloves, high yielding and resistance to diseases

and pests. The new varieties should be thicker as that of China. It is because of this thickness that Chinese garlic is in huge demand in the international market.

(2) Need to provide proper marketing facilities along with transportation and storage facilities to the farmers.

(3) As and when surplus crop appears, there is the need to channelise it into the production of garlic powder by improved techniques. This can minimise the wastes. This will also assist in regulating the market rates, especially in the glut season.

(4) All undersized or cull but healthy bulbs, which normally fetch lower price, could be conveniently utilized in the manufacture of garlic powder.

(5) Garlic powder converted into suitable tablets or packed into capsules should find a ready market in India, in the face of costly synthetic tablets claiming anti-bacterial activity.

(6) Need to develop export worthy varieties and the promotion of export to European countries.

(7) Need to expand the processing activities.

The expected demand for Garlic by 2020 AD will be 7 lakh tonnes. This target can be met out by expanding the area under garlic in non-traditional but potential tracts. The scope for increasing the productivity is also there by adoption of these efforts.

4
Fruit Vegetables

This chapter discusses on Solanaceous fruits and Okra. The Solanaceous fruits group includes three important vegetables, viz., tomato, egg-plant or brinjal and chilli. India is the leading producer of all these vegetables in the world. It ranks second in brinjal and sixth in tomato. Okra constitutes a major item in our vegetable exports. Brinjal is a native of India and has been in cultivation since ancient times. All of these fruit vegtables are grown throughout India and are the most common vegetables used by the Indians.

All of these vegetables are having special nutritive values and are used in medicines too. Over the years, several improvements have been made in the cultivation of these vegetables in India with the introduciton of hybrids and modern technologies. However, there are several problems in their cultivation as well as in the post-harvesting activities. These problems demand proper attention and a sound strategy. In this chapter, a detailed discussion is made on these vegetables in connection with past, present and future aspects.

TOMATO

Tomato (*Lycopersieon esculentum*) is native of Peru-Ecudor area, from which it spread northward in pre-Columbian times to Mexico where it was first cultivated. The Spanish explorers carried the plant to Southern Europe where it was eaten for a long time before it was utilized by the people of Northern Europe and the United States. For many years, it was considered to be poisonous, and was grown only for ornamental purposes under the names 'Tomatl', 'love apple'. Today it ranks next to potatoes and sweet potatoes in importance.

Tomatoes are coarse, branching, erect or trailing herbs, with a true berry for a fruit. They differ greatly in habit depending on the environmental relations. Tomato is a warm-season crop, the best soil for tomato is rich loam, with a little sand in the upper layer, and a good clay in the sub-soil.

Values and Uses

Abundantly rich in Vitamins, minerals and organic acids the tomato fruits contain 3-5 per cent total sugar, 15-30 mg/100 gm ascorbic acid, 7.5-10.5 mg/100 ml titratable acidity, 2.4 g/100 g dry fruit weight of minerals and 25.50 mg/100 g of lycopene. Besides being popular for soup and ketchup, the tomato fruits are essentially used in cooking curry vegetables. The nutitive value of tomato varies in different varieties and also depeds upon the environment in which it is cultivated. As a whole, fresh ripe fruits are refreshing and appetizing and are consumed raw in *salads* or after cooking. Unripe fruits are caned. Tomatoes are also consumed in the form of juice, paste, ketchup, sauce, soup and powder.

World Production

In the world, tomato is cultivated in almost all regions. In terms of area under this head, the Asian region occupies nearly half of the total. The total area under tomato in 1998-99 was 3254 thousand hectares. As far as the area under this crop is concerned, China stands first followed by India, U.S.A., Turkey, Egypt in the world. In terms of production, China is the leading producer followed by U.S.A., Turkey, Egypt, Italy, India and others. The total production during the period 1998-99 was 90,360 metric tonnes which was 90,468 metric tonnes in the previous year. (Table 4.1).

Area and Production in India

Tomato is one of the most popular vegetables grown successfully throughout Inda. The total area under this crop at present is 485,520 hectares and the production is about 5,300 metric tonnes. As far as area under tomato in India is concerned, Orissa stands first followed by Bihar, Andhra Pradesh, Karnataka, Uttar Pradesh and others and in terms of production Uttar Pradesh ranks first followed by Bihar, Karnataka, Maharashtra, Orissa, Andhra Pradesh and others as can be seen from Table 4.2. This table shows the share of the major producing states in terms of both area and production in India. Today tomato has become a very popular fruit vegetable crop and occupies second place in terms of acreage next to potato. The production of tomatoes in India got a boost when Pepsi promoted its cultivation in Punjab with its high yielding seeds.

Table 4.1 : World Production of Tomatoes

(in million tonnes)

Year	Production
1979-81*	53.79
1985	60.83
1989-91⁺	75.02
1991	75.25
1992	72.36
1993	70.62
1994	81.37
1995	85.32
1996	91.56
1997	90.46
1998	90.36

+ Triennium Average.
Source : FAO.

Varieties

The tomato can be grown almost throghout the year in India. The number of crops grown varies from region to region. In the northern India generally two crops, autumn and spring crops, while on the hills, only one crop is taken. Several varieties are grown in India. The important ones are 'Pusa Ruby', 'Pusa Early Dwarf', Pusa Gaurav', 'Punjab Chhuhara' 'Selections 120 and 4, 'HS101, 'Arka Vikas', all of these are the recommended by IARI. While 'Rupali, 'Rashmi', 'Naveen', 'Mangala', 'Vaishali', 'Karnataka', 'IAHS 88-1', 'Sheetal' have been among hybrids all from the indo American Hybrid Seeds Company, Bagalore. They have built-in specific qualities like early yield, high total yield, resistance and toughness for long distance transport. Now-a-days, wide range of high yielding hybrid varieties to tomato are available

Table 4.2 : State-wise Output and Yield of Tomato

(Area, Hectares, Output, M.T. : Yield M.T./ha)

State	1993-94			1994-95			1995-96		
	Area	Output	Yield	Area	Output	Yield	Area	Output	Yield
Andhra Pradesh	47,156	4,71,560	10.00	46,907	4,69,070	10.00	47,723	4,77,230	10.00
Arunachal Pradesh	250	680	2.72	250	680	2.72	250	680	2.72
Assam	12,840	2,50,380	19.50	13,100	2,46,280	18.80	13,150	2,83,800	21.58
Bihar	55,610	11,12,200	20.00	57,900	11,58,000	20.00	54.400	11,28,000	20.74
Gujarat	7,000	1,05,000	15.00	7,500	1,12,500	15.00	14,057	1,83,882	13.08
Haryana	5,400	99,700	18.46	6,400	1,25,000	19.53	7,600	1,45,200	19.11
Himachal Pradesh	2,366	59,000	24.94	2,420	65,457	27.04	2,570	79,057	30.75
Karnataka	32,073	8,01,825	25.00	39,207	9,80,175	25.00	36,516	9,12,900	25.00
Madhya Pradesh	23,389	3,50,000	14.96	23,840	3,57,000	14.97	25,032	3,75,000	14.98
Maharashtra	43,290	3,82,274	8.83	29,326	4,46,791	15.24	28,752	5,20,147	18.09
Manipur	396	3,170	8.01	400	3,200	8.00	406	3,250	8.00
Orissa	50,2000	5,82,300	11.60	54,300	6,16,300	11.35	55,929	6,47,115	11.57
Meghalaya	395	2,469	6.25	698	9,367	13.42	700	9,800	14.00
Mizoram	181	751	4.15	148	532	3.59	219	795	3.63
Nagaland	152	1,946	12.80	528	6,336	12.00	552	6,624	12.00
Punjab	5,600	1,35,486	24.19	5,600	1,35,486	24.19	5,750	1,39,093	24.19
Rajasthan	12,892	60,296	4.68	15,792	53,911	3.41	14,232	52,525	3.69
Tamil Nadu	20,445	2,65,580	12.99	21,055	2,32,430	11.04	21,055	2,32,430	11.04
U.P. (Hills)	5,803	39,203	6.76	6,054	38,677	6.39	6,032	37,727	6.25
U.P. (Plains)	4,415	25,667	6.19	3,937	24,377	6.19	4,244	26,283	6.19
West Bengal	14,109	1,45,755	10.33	14,109	1,45,755	10.33	14,109	1,45,755	10.33
Daman & Diu	25	425	17.00	25	425	17.00	25	425	17.00
Delhi	1,054	24,124	22.89	1,490	21,080	14.15	1,595	22,126	13.87
Pondicherry	265	4,781	18.04	298	5,960	20.00	293	5,567	19.00
Total	3,45,943	49,33,980	14.26	3,51,787	52,61,277	14.96	3,55,684	54,41,969	15.70

Source : National Horticultural Board.

in the market on large scale. Recently eleven improved varieties including 4F, hybrids have been released for cultivation. Three open polinated varieties namely Arka Saurabh, Arka Ahuti and Arka Ashish suitable for processing yielded 35 to 45 tonnes/ ha under non-stated conditions. Four high yielding varieties including two F1 hybrids with resistance to bacterial wilt have been released for fresh market. Arka Shresta and Arka Abhijit yields about 45-55 tonnes per hectare.

As far as hybrids varieties of tomato are concerned, there are two categories viz., (1) Determinate and (ii) Indeterminate. In the case of determinate varieties, the plants grow to normal height, bear fruits and after specific period plant growth stops, the fruits mature and the life cycle ends. While in the case of indeterminate varieties, the fruits bearing takes place during the plant growth. Both the plant growth and fruit bearing are continues simultaneously till the end of the life cycle of the plant. Now, F1 hybrids of tomatoes are being widely used in Inda in all the important tomato growing states and it covers nearly 32 per cent of the total area under tomato. At present, Karnataka, Maharashtra, Punjab, Haryana, West Bengal, Orissa and Uttar Pradesh account for maximum area under hybrid tomatoes.

Production of the Hybrid Seeds of Tomato in India

In India, the laboratories contributing to the improvement of tomato crop are the India Institute of Horticultural Research, Hessaraghatta in Karnataka; IARI, New Delhi, Punjabn Agricultural University, Ludhiana, Haryana Agricultural University, Hissar and Jawaharlal Nehru Krishi Vishwa Vidyalaya. Jabalpur. Apart for these, Mahyco at Thaina, Maharashtra, Suttons, Century Seeds, Indo-American Seeds Company in Karnataka, Sandoz (India) Ltd., Sungoe, Nijjan Agro Foods Limited and SPIC, Nath Seeds and Ankur Seeds are also engaged in the production and marketing of hybrid seeds in India.

Harvesting

The stage of maturity of which tomatoes are to be harvested depends on the purpose for which they are cultivated and the location and distance over which they are to be marketed. As a whole, according to its use, they are harvested in the following stages :

(a) Green Stage

Tomatoes in this stage are harvested mainly for the purpose of maketing these at distant markets.

(b) Pink Stage

During this stage, the fruits become red, however, they are not fully ripe. Harvesting of these are done mainly to cater the demand in the local marketing.

(c) Ripe Stage

During this stage, the surface of the fruits becomes red and the softening of fruits begins and they are harvested for domestic consumption or for table use.

(d) Full Ripe Stage

At this stage, maximum colour development takes place and they fall soft to touch. These fruits are normally used within a day of picking and are consumed or used for canning and pickling.

Tomato Products

(1) Canned Tomatoes

Ripe tomatoes of medium size, regular shape, uniform red colour solid meat and of good lavour are selected for canning.

They are washed in boiling water or steam for 2-3 minutes, peeled and canned.

(2) Tomato Juice

Ripe Tomatoes with bright red colour and high acidity are used for the extraction of juice. Tomato juice is higly esteemed as an appetizing and nourishing beverage. It is sometimes seasoned to produce cocktail known as Tomato juice cocktail. Tomato juice powder is prepared by dehydrating the juice.

(3) Tomato Ketchup

It is prepared from ripe tomatoes of deep red colour by cooking the pulp in kettles with spices. Cooking is continues till the desired consistency is obtained. Vinegar, salt, sugar and sometimes pectin are added to this.

(4) Tomato Soup

For the preparation of tomato soup, the fruit pulp is partly neutralized by adding sodium bicarbonate solution and then concentrated in a pan, spices, arrowroot and butter or cream are added. When the desired consistency is obtained, salt and sugar are added and the mass boiled for another two minutes.

Marketing

There is no organised marketing system for tomato in India. A major portion of the product is still handled by the middlemen. For the market, the grading of tomatoes is being done to some extent for specialised urban markets. The Bureau of Indian Standards has specifies four grades namely Super A, Super, Fancy and Commercial.

Exports

India exports small quantities of tomatoes and its products mainly to Bangladesh, Nepal, Sri Lanka, Pakistan, Scychelles, Malaysia, UAE, U.S.A. and Mexico. Its exports went up from 187 tonnes valued at Rs. 6.09 lakhs in 1993-94 to 1,072 tonnes valued at Rs. 62.94 lakhs in 1994-95, however, it has more or less remained same at 646 tonnes valued at Rs. 36.11 lakhs in 1995-96 and 690 tonnes valued at Rs. 41.96 lakhs in 1996-97, but in 1997-98 the exports rose to 863 tonnes and its value was Rs. 41.38 lakhs (Table 4.3).

The Tomato Industry in the World

It is a cyclic industry where every 4-5 years there is a dip in the market prices, may be due to over production or over estimates. It is a risk-prone inudstry and is mainly depending upon the nature. The U.S. is the biggest producer of processing tomatoes in the world. California alone contributes 85 per cent of the output. U.S.A. is also the highest importer of tomato products ike tomato paste, puree, diced tomato, chopped tomato in puree, dried tomato, whole-peel tomato in puree, slasa and subgady sauce.

Table 4.3 : Country-wise Export of Fresh or Chillied Tomatoes from India 1997-98

(Volume in Kg. Value in Rs.)

Country	Volume	Value
Bangladesh	353480	2398309
Nepal	326925	569323
Sri Lanka	69079	472566
Pakistan	44503	237412
Scychelles	23600	211797
Malaysia	22000	6C051
U.A.E.	17910	141984
Canada	3050	30069
U.S.A.	2000	16845

Source : DGCIS Ministry of Commerce.

The major producers of tomato products in the world are the U.S.A., Italy, Spain, Portugal and Turkey. Of the total production of tomato in the world, one-third is processed as tomato paste (the base for soup and ketchup), juice, whole peeled and diced tomatoes and the remaining two-thirds of the fruit is consumed fresh. In recent years, the demand for Toamto paste is increasing in the international market because of several reasons. They are : economical substitute for fresh Tomatoes during the off-season, aseptic packaging has increased the shelf-life of the processed products pasta and pizza are relished with sauce and ketchup, and pizza culture is spreading dramatically; tomato-based soup is becomig popular on religious occasion in the Islamic countries, while North American prefer red colour in their meals, toamto paste or its soup and gravy are ideal.

The main exporters of these products on the world are Italy, Greece, Turkey, Portugal, Spain, Chile, Argentina, Israel, Hungary, Bulgaria, Romania, Mexico, Brazil, Peru, Libya, Tunisia and Taiwan while the major markets for these products or the importers are the U.K., Germany, Greece, Belgium, the Netherland, Denmark, Norway, Sweden, Switzerland, Australia, Yugoslavia, Finland, Russia, Czechoslovakia, the U.S.A., Canada, Algeria, Libya, Egypt,

Sweden, Zambia, South Africa, Iraq, Lebanon, Jordan, Kuwait, Malaysia, Indonesia, Japan, Korea and the West-Asian countries.

In India, more than eleven units are engaged in the production of tomato paste of international quality and they produce more than 250 tonnes of paste a day. The main units are The Clean Foods Corporation and Sunsip in Andhra Pradesh; Kissan Products and Bulgaria Foods in Karnataka; Bihar Agro Fruit and Vegetable Processing Corporation in Bihar; Punjab Agro Foods, Pepsi Foods and Nijjar Agro Foods in Punjab; Regency Foods in Rajasthan; Kissan Fruit Products in Uttar Pradesh and NAFED in Tamil Nadu.

Problems

(1) Problems of diseases and insect pests have minimised the scope for an increase in the production and productivity.

(2) Non-availability of F1 hybrid seeds at lower price.

(3) Non-availability of procesing varieties.

(4) Absence of proper marketing systems and lack of sufficient storage facilities.

(5) Lack of incentives for processing and experts.

Since, there is a vast scope for the development of this sector, efforts are needed to overcome the prevailing problems and to boost the production and productivity. Along with these, there is also the need to promote the production of value based products of tomato. In this regard, encouragement is neded from the small scale organisations to establish tomato based processing units with necessary low cost technology in growing areas. This will push up our exports on the one hand and on the other grower's income also.

BRINJAL

The eggplant or Brinjal or Aubergine (*Solanum melongena*) is a native of India and has been in cultivation for a long time. It is widely grown in the warmer regions of both hemisphere, especially in the West Indies and Southern United States. The plant is an erect branching herb or small shrub and the fruit is large, ovoid, whitish or purple berry.

Values and Uses

Nutritionally, Brinjal can be compared to Tomato. The edible portion of Brinjal per 100 g contains : 92.7 g of moisture, 1.4 g of protein, 0.3g of fat, 0.3 g of mineral, 1.3 g of carbohydrates, 4 g of calcium etc. The fruits of Brinjal are an excellent remedy for those suffering from liver troubles. The green leaves of the Brinjal plant' are the mian sources of supply ot antiscorbutic Vitamin C. Its seeds are used as a stimulant. White Brinjal is said to be good for diabetic patients.

World Production

The total area under this crop in the world in 1999 was 1236 thousand hectares and the production was 20147 thousand M.T. China is the largest producer followed by India as can be seen from Table 4.4.

Table 4.4 : Area and Production of Egg-plants in Major Producing Countries in the World

Country	1997		1998		1999	
	Area	Prodn.	Area	Prodn.	Area	Prodn.
China	551	10026	551	10026	551	10026
India	420	6000	420	6000	420	6000
Turkey	33	850	33	850	33	850
Egypt	28	555	29	560	29	560
Japan	15	490	15	490	15	490
World total (including others)	1231	20088	1236	20164	1236	20147

Source : FAO.

Area and Production of Brinjal in India

Brinjal is one of the common vegetables grown throughout the country and the total area under this crop may be above 3,20,000 hectares, since, there is no reliable statistics on its area in our country, the above figure is just an estimation. Now, in our country, it is considered as one of the major commercial crops. As per the estimation of FAO, the total area under Brinjal in India is in 420 thousand hectares in 1999 and the production was 6000 thousand tonnes which is considered to be the highest volume out of the total vegetable production in India (Table 4.5)

Table 4.5 : Area and Production of Brinjal in India

Year	Area ('000 ha)	Production (in '000 M.T.)
1990	295	3000
1991	300	3124
1992	303	3150
1993	305	3200
1994	308	3150
1995	310	3300
1996	320	3400
1997	420	6000
1998	420	6000
1999	420	6000

Source : FAO.

Varieties

Brinjal has three main varieties viz., the round or egg-shaped, the long and the dwarf. According to the eight agro-climatic zones namely (1) Humid Western Himalayan region; (2) Humid Bengal-Assam Basin; (3) Humid Eastern Himalayan and Bay Islands; (4) Sub-humid Sutlej-Ganga Alluvial Plains; (6) Arid Western Plains; (7) Semi-Arid Lava plateau and Central higlands; and (8) Humid to semi-arid Western Ghats and Karnataka, so far 14 varieties have been identified. They are :

(1) Pusa Purple Long, which is an old variety and was found suitable for zones, 4, 6, 7 and 8. The fruits of this variety is long, purple, glossy and tender. This is suitable particularly in the Northern part and can be cultivated both as a summer and autumn crop.

(2) Pusa purple cluster, is a medium-early variety and was found suitable to zones 1, 5, 6 and 7. The fruits of this are log, deep purple in colour and borne in clusters of 4-9. It is suitable for southern and northern hills and is moderately resistant to bacterial wilt.

(3) pH4 is suitable for 4th and 6th zones. The fruits of this are medium to long and thin with dark purple colour.

(4) Pusa Kranti is best suited to 4th zone. This gives long fruit with dark purple and the seeds in this are lower.

Pusa Kranti can be cultivated both in the spring and autumn seasons.

(5) Pant Samrat is resistant to Phomopsis blight and blight and bacterial-wilt when the crop is grown in the field. This is identified for 4,5 and 6 zones. The fruits of this are attractive.

(6) Kt4 has medium-sized plants and was identified for 1, 4 and 8 zones. The fruits of it are borne in clusters and are cylindrical with a purple colour.

(7) ARV 2C was identified in 1987 for 1, 4 and 8 zones.

(8) Azad Kranti was identified in 1983 and is suitable for 1 and 3 zones.

(9) K 202-9 was identified in 1987 for zones 6 and 8.

(10) Pant Rituraj is best suitable for 4, 6, 7 and 8 zones. The fruits of this are round with an attractive purple colour. This fruit is soft, less seeded and endowed with good flavour.

(11) T3 was developed in 1975 for 4, 6 and 8 zones. The fruits of this are round, light purple with whitish green colour at stigmatic end.

(12) Jamuni Gola was identified in 1975 for 5 and 6 zones. This is an early maturing variety.

(13) Arka Kusumakar is from IIHR. This was identified in 1981 for 4, 6 and 7 zones. The fruit of this is small, borne is clusters of 5-7, good texture and cooking quality.

(14) Arka Navneet was identified in 1981 for the cultivation in 4, 6, 7 and 8 zones. The fruits of this are oval and free from bitterness.

Apart from these, several new varities entered recently. The Punjab Agricultural University identified some promising cultivars namely, Punjab Barsati, Sada Bahar Baingan, Pb8 etc., At U.P. the promising cultivars evolved are PBr 61, Sel 1, PBr 91-1 and PBr 91-2.

In Tamil Nadu cluster white, CO 1, SM 609, Annamali etc., have found promising cultivars Makra, Elokeshi etc. are popular in West Bengal. Nurki long, Round etc., are poplar in Madhya Pradesh. Improved Muktakeshi 1 is popular in Bihar and in Karnataka Arka Kusumakar, Arka Sheel and Arka Shirish are popular.

As Brinjal is warm-season crop, it can be grown in all types of soils varying from light sandy to heavy clay. However, silt-loam and clay-loam soils are preferred for its cultivation.

Harvesting

Brinjal is harvested when it matures. However, it is harvested before it fully ripens, here care is taken to attain a good size and colour. Harvesting is normally taking place when the surface of the fruit is bright and glossy in appearance.

Demand

The Brinjal fruits may be white, yellow, brown, green, black, pink, purple and striped in colour, and in shape they may be round, long elongated, pear-shaped and oblong depending upon the cultivars. The long green varieties are preferred in Bihar and Karnataka and the round, the green in Orissa. While in North India, pinkish purple or violet and black colour varieties are demanded. As far as the yellow, brown and white coloured Brinjal fruits are concerned, the demand is lower.

Marketing

For the market, Birnjal is graded into three grades viz., Super, Fancy and Commercial. As far as marketing or this is concerned, there is no organised marketing system. Bulk of the trade is still traditional which consists of the involvement of middlemen, retail and wholesale traders.

Prolems

(1) Non-availability of cold storage facilities for storing Brinjal at the production and marketing centres.

(2) Poor marketing systems.

(3) The prevalence of a number of diseases like damping off, Phomopsis blight, Bacterial wilt etc.

(4) Insects like Epilachna beetle, shot and fruit borer, red mite etc.

(5) Non-availability of disease free and resistant varieties to the farmers.

(6) Non-availability of proper training for the growers on pre- and post-harvesting aspects.

As India stands first in the production of Brinjal in the world and the export prospects are bright, there is an urgent need to overcome these problems with a proper plan. Efforts are also needed to increase the productivity and production in the country. In this regard, an expansion in its area under hybrid varieties is needed, since the total area under hybrid varieties is around 18 per cent of the total cropped area.

CHILLIES

Chilliea are the dried ripe fruits of the spices of genus capsicum. They are also called red peppers or capsicums and they constitute an important, well-known commercial crop used both as a condiment or culinary supplement and as a vegetable. Chilli was not known to Indians about 400 years ago, since this crop was first introduced into India by the Portuguese towards the end of the 15th century. Its cultivation became popular in the 17th century. Chilli is actually reported to be native of South America and its cultivation was known to the natives of Peru since prehistoric times.

Values and Uses

Chillies are rich in Vitamins, especially Vitamin A and C. Green chillies per 100 g of edible matter consists of 85.7g moisture, 2.9g Protein, 0.6g Fat, 1.0g Minerals, 6.8 g fibre, 29 g Calories etc. Chillies when eaten fresh with *salads*, they, being rich in Vitamins C and A, serve as a good vitamin supplement in addition to their being an appetizer. Green chillies are more nutritious than ripe, dried chillies or chilli powder.

Paprika, Red pepper, Kashmiri Mirch are mild in pungency and are used to colour, flavour and garnish dishes, while Simla

Mirch which is not pungent is used as curried vegetable, is specially rich in Vitamin C.

Dry chilli is extensively used as a spice in all types of curried dishes in India and abroad. Bird chilli. is used in making hot sauces such as pepper sauce.

Capsicum preparations are used as counter-irritants in lumbago, neuralgia and rheumatic disorders.

Area, Production and Yield of Chillies in India

In India, chilli is grown in almost all states. Andhra Pradesh has the highest area and production followed by Karnataka, Maharashtra, Punjab, Rajasthan, West Bengal and others. Productivity of chilli is the highest in Andhra Pradesh followed by Punjab, Rajasthan and Karnataka as can be seen from Table 4.6. The All India area under this crop during 1997-98 was at 831.5 thousand hectares and the production was at 821.8 thousand tonnes which was less than the are in 1996-97 that of 944.2 thousand hectares and in terms of production, it was 1066.4 thousand tonnes. The fall in the production as well as area under chillies was reported in states like Andhra Pradesh, Karnataka, Rajasthan and Tamil Nadu during the period. However, it went upto 892.2 tousand hectares with a production of 921.3 thousand tonnes in 1998-99. Table 4.7 gives data for the trend in area under production of chillies in India over the years.

Varieties

The varieties under cultivation differ in the size, shape, colour and pungency of the fruits.

The varieties of chillies are broadly divided into two groups namely (i) the long pungent type, including pickling type, used as a spice and (ii) the bell-shaped, non-pungent or mild and thick fleshed type, popularly known as 'Simla Mirch' which is commonly used as a curried vegetable. 'Paprika' also belongs to the mild group. Most of the bigger red-coloured fruits cultivated and marketed the world over, inlcuding the chillies, paprika and capsicum belong to the species *Capsicum annum*; the highly pungent small variety belongs to the species *Capsicum fruitescens*. The early

Marketing of Vegetables in India

Table 4.6 : State-wise Area, Production and Yield of
Chillies in India 1998-99
(Area : '000 ha, Production : '000 tonnes, Yield: Kg/ha)

State	Area	Production	Yield
Andhra Pradesh	225.9	403.0	1785
Arunachal Pradesh	1.2	1.6	1333
Assam	14.7	9.7	660
Bihar	6.1	4.7	770
Gujarat	18.1	18.2	1006
Haryana	1.4	1.5	1071
Himachal Pradesh	1.1	0.3	273
Karnataka	178.4	142.6	799
Madhya Pradesh	48.1	19.7	410
Maharashtra	101.1	57.7	571
Manipur	8.8	5.3	602
Meghalaya	1.8	1.1	611
Mizoram	2.8	3.3	1179
Nagaland	1.2	9.6	8000
Orissa	90.1	76.6	850
Punjab	4.7	8.0	1702
Rajasthan	33.5	49.2	1469
Tamil Nadu	66.0	39.7	602
Tripura	2.0	1.2	600
Uttar Pradesh	19.6	15.5	791
West Bengal	64.4	51.3	797
Total (Including Others)	892.2	921.3	1033

Source : D of E&S New Delhi.

Table 4.7 : Area and Production of Chillies

Year	Area (in ha)	Production (in tonnes)	Yield (Kg/ha)
1970-71	783,400	520,400	664
1975-76	739,800	526,100	711
1980-81	834,800	509,100	610
1985-86	904,100	877,400	970
1990-91	816,200	719,000	881
1995-96	883,700	809,400	916
1996-97	944,200	1066,400	1129
1997-98	831,500	821,800	988
1998-99	892,200	921,300	1033

Source : D of E&S.

botanists selected the term Capsicum to designate the genus belonging to the family of Solanaceae.

The term 'Paprika' is genrally used for non-pungnet (sweet) red capsicum powder. Red Chile (pungent) and Paprika are dehydrated and sold as whole fruits or grounded into powder. In the United States, Paprika is made from the New Mexican type chilli, whereas in Europe, Paprika is made from two main fruit types (i) A round fruit about the size of a peach called Spanish or Moroccan paprika. In fact, the Hungarian word for plants in the genus Capsicum is "Paprika" and it may be pungent or non-pungent.

The perennial are known as 'bird chillies' belong to C. *frutescens* Linn. The bird chillies are very pungent, short-lived and grow for 2 or 3 years. They are enlisted in the *British Pharmacopoeia* and find maximum use in pharmaceuticals.

The different varieties available for cultivation are 'NP 41' a high yielding pungent chilli and 'NP 46', another pungent chilli resistant to thrips. 'Hybrid 5-1-5' is high yielding and suitable for the nproduction of green chillies. Among the non-pungent vegetable types, there are two American varieties viz., 'World Beater' and 'Bell pepper' and one Russian variety, 'RH 49' are high yielding. In Andhra Prdesh, the improved varieties 'G1, G2, G3, G4, and G5 and four cultures, 'X200', 'Ca 960', 'X196' and 'X197' are high yielding and are fast spreading in the state. The variety 'G1' is high yielding tolerant to thrips, has a persistent Calyx and is higly suited for export.

Harvesting

In North India, it is harvested in about 60-75 days while in South-90-105 days. For vegetable purposes, it is harvested while they are still green but full grown.

Marketing

The crop becomes ready for harvesting in about 3-4 months after planting. After picking the fruits, they are dried in the sun for 4-5 days, and are grades for size and colour before making. Unripe chillies are sometimes boiled and dried for domestic consumption. Commercially there are various grades, such as the first sort, thesecond sort, mixture etc. Grades such as special, medium and fair are also adopted. Good fruit length, shining red colour, high pungency and strong attachment of the calyx are the important factors which the merchant consider for fetching a high price.

The major chilli marketing centres in India are Nasik, Ahmednagar, Sholapur, Aurangabad, Nanded, Amaravathi, Lasalgaon in Maharashtra, Guntur, Warrangal, Hyderabad, Cuddapah, Vijayawada, Rajahamundri and Nellore in Andhra Pradesh, Dharwad, Mysore, Hassan, Bangalore, Bellary, Ranibennur, Hubli and Byadagi in Karnataka, Pollachi, Ramnad, Madurai, Trichi, Theni, Dindigal, Virudu Nagar and Sathu in Tamil Nadu.

The marketing of chilli remains a major constraint since the commission agents still take a major share of consumer's price.

Export

India exports fresh chilli, chilli seed, dry chilli, chilli powder, *Capsicum genus*, chilli oil and Oleoresin to U.S.A., U.K. UAE, Saudi Arabia, Singapore, Sri Lanka, Bangladesh and other Asian countries. Export of chilli during 1998-99 has shown substantial increase both in terms of quantity and value. The export during the year was 61,253 tonnes value at Rs. 216.61 crores as against 51,779 tonnes valued at Rs. 158.90 crores in 1997-98.

The export of chilli to Sri Lanka has increased from 11881 tonnes in 1997-98 to 19796 tonnes in 1998-99. Export to UAE has also increased from 2523 tonnes to 3467 tonnes in the same period. However, export to USA has declined from 12057 tonnes to 7555 tonnes; to UK it has decreased from 2084 tonnes to 1401 tonnes.

As far as the export of fresh chilli is concerned, the total volume of exports during the period 1998-99 was 11926.57 tonnes valued at Rs. 4950.29 lakhs while it was 14644.99 tonnes in 1997-98 valued at Rs. 5687.62 lakhs. Table 4.8 gives data on the exports of chillies since 1993-94 up to 1998-99.

Table 4.8 : Export of Chillies from India : 1993-94 to 1998-99

(Volume in M.T. Value in Rs. Lakhs)

Item	1993-94		1994-95		1995-96		1996-97		1997-98		1998-99	
	Volume	Value	Volume	Value	Volume	Value	Volume	Value	Volume	Value	Volume	Value
Chilli dry	22980.6	5243.61	12813.84	3579.85	43969.44	14973.38	32711.69	13257.70	36024.45	9939.22	48273.33	16402.21
Chilli seed	-	-	-	-	108.97	28.00	299.22	90.35	600.01	137.31	792.05	223.37
Chilli Fresh	293.3	40.95	106.90	16.86	141.42	17.66	961.14	85.11	303.11	30.10	97.19	20.10
Chilli powder	7,048.0	1899.99	7102.06	2096.12	11900.98	4515.09	15904.14	6623.90	14644.99	5687.62	11926.57	4950.29
Capsicum genus	46.7	19 01	34.90	12.35	-	-	132.61	63.90	95.81	39.56	37.64	20.31
Other Pimenta	7.7	10.37	28.66	6.44	44.00	12.04	52.21	24.20	110.95	56.21	126.02	44.83
Chilli Oil	6.8	25.10	1.43	19.53	0.25	1.31	0.01	0.08	1.15	23.40	0.22	3.86
Paprika Oil	-	-	-	-	3.60	14.44	-	-	5.48	33.88	-	-
Capsicum Oil	-	-	0.42	16.05	0.35	3.54	0.67	10.69	1.54	19.22	-	-
Chilli Oleoresin	71.9	778.72	45.94	498.26	14.28	126.77	11.71	165.94	105.00	1063.27	27.52	229.45
Paprika Oleoresin	59.2	484.22	82.43	774.34	157.39	1734.07	261.51	3187.28	487.98	5818.04	731.52	7865.21
Capsicum Oeloresin	183.9	1115.52	227.11	1324.73	195.26	1337.79	268.52	2163.53	259.12	2008.07	290.34	1986.55

Source : DGCI & S.

Problems

(1) Fluctuation, stagnation or low yield in chillies is common in India. This has been due to the cultivation of disease-susceptible varieties, especially the virus susceptible chillies. A large number of production constraints limit chilli production in the country. The constraints are – Incidence of leaf curl and bacterial wild, lack of varieties high in Capsaicin, Oleoresin and capsanthin and absence of hybrid technology. The yield in India is much lower when compared to Indonesia, Brazil and Malaysia.

(2) Among the produvction constraints, incidence of leaf curl and bacterial wilt are major problems. A large number of complex diseases like pepper virus complex, leaf curl complex, Rapid decline, Mosaic complex and Moria disease limit chilli production in India. Various studies on chillies show that leaf curl leads to 100 per cent loss and fruit rot by Alternaria leads to 50 per cent of loss.

(3) Marketing of chillie is a major constraint. The major chilli marketing centres in the state of Maharashtra, Andhra Pradesh, Karnataka and Tamil Nadu show that the commission agents still take a major share of consumer's price.

(4) Cost of cultivation in chilli is going up in recent years due to high cost of labour and plant protection chemicals.

(5) The cost of transportation is ever increasing.

(6) Non-availability of proper curing methods.

(7) The main reason for the declining exports is high prices as compared to Pakistan and China who are the main competitors for India.

(8) Our chilli exports suffer because of adulteration and poor packing.

Prospects

India has immense potentiality to grow chillies. The world demand is also going up for chillies, which is more than 40,000 tonnes. The volume of exports of chillies, chilli powder and

Oleoresin of chillies is too small in relaiton to the world production and demand. As India has been traditionaly exporting these, there is vast scope for increasing this. In ths regard, there is an urgent need for a long term strategy, which should contain the following impotant aspects :

(1) There is the need to identify varieties as resistant to varius diseases, especially to leaf curl complex.

(2) There is the need to come up with a set of agronomic practices to increase the productivity. In this regard, control measures against dieback, fruit rot, powdery mildew and bacterial leaf spot has to be taken. Again control measures have to be worked out against major pests of chilli like podborer and pest complex.

(3) Need to identify high yielding and processing oriented varieties which should contain more of Oleoresin, colour and capsaicin. This can meet the requirement of the chilli growers, traders, consumers and industrialists.

(4) So as to overcome the problem of marketing and that of commission agents, the only solution is co-operative type of marketing which is yet to take a start in chilli.

(5) For improving exports improved flavour, pungency, vitamins and quality of Oleoresin can build into the chilli varieties.

(6) So as to minimise waste as and when production is more, curing should be done by the "netsack method" which is practised in Hungary.

(7) The world trade in Paprika Oleoresin is showing an upward trend in recent years, so, there is an imperative need to develop high yielding Paprika like chillies mildly pungent and having high colour value as there is great demand for such varieties in the international market.

Above all, agro techniques, seed-production techniques and development of varieties resistant to biotic and aboiotic stress and ideal processing methods in this crop have to be standardised. Then ony India can hope to gain the major external markets like West Asia, North Africa, Russia, EECs, USA, Canada, Japan and Australia for chilli and its products.

OKRA

Okra (*Hibiscus esculentus*) is a native of tropical Africa. It was cultivated by the Europeans as early as in the thirteenth century and has been introduced into warm tropics and subtropics. The plant is a stout annual, much resembling cotton in its habit. It is an annual vegetable crop grown in the tropics of the world. Okra being a warm season crop, requires a long warm growing season. It can be grown on all kinds of soils, however, to get best results, it requires a friable well-manured soil.

Values and Uses

The vitamin and mineral reich Okra is a prized vegetable in the Indian sub-continent. It has an average nutritive value of 3.21 which is higher than tomato, brinjal, pumpkins and ash gourd. It contains 27g carbohydrates, 2.2 g proteins, 0.29g fat, 90mg calcium, 50mg phosphorus, 15mg Iron and 16mg Vitamin C per 100g of edible portion. The seeds contain 18-20 per cent oil and 20-23 per cent crude protein.

Soups and stews of Okra are popular dishes in India. The seeds, when ripe are sometimes roasted and used as a substitute for coffee. The roots and stems are used for clarification of sugarcane juice before it is converted into jaggery and brown sugar. The crop is used in paper industry and the stem is used for the extraction of fibre. Roasted and ground seeds find use as fullers in many beverges. The medicinal values of Okra are associated with genitourinary disorders, spermatorrhoea and chronic dysentery.

Area, Production and Productivity in India

Okra is also called 'Ladyfinger'. I India, it is grown throughout except in the mountain region. The major producers are Uttar Pradesh, Bihar, Orissa, West Bengal, Assam, Andhra Pradesh and Karnataka. As far as area under Okra in India is concerned, Uttar Pradesh has an area of 1.73 lakh hectares, Orissa 0.83 lakh hectares followed by Bihar, West Bengal and others. In terms of production, it is 15.1 lakh tonnes in Uttar Pradesh, 10.39 lakh tonnes in Bihar and 7.21 lakh tonnes in Orissa. The average productivity in India is 9.36 metric tonnes per hectares, the highest is in Bihar that of 14 metric tonnes per hectare followed by Himachal Pradesh which is 12.5 tonnes per hectare (Table 4.9). As a whole, in the plains of northern India, normally two crops are raised viz., one in the early spring for the summer crop and the other in late summer. While in the southern parts of India, a winter crop is also raised.

The total area under this crop moved in the positive direction over the years in India and that in terms of production too. In 1992-93, the total area udner this crop wss 2.81 lakh hectares and the production was 24.87 lakh tonnes and in 1995-96, it was 40.33 lakh tonnes from an area of 4.31 lakh hectares.

Varieties

There are a large number of varieties of Okra and they may be classified as tall, medium and dwarf. They are further classified according to the quality of the pod, some are with prominent ridges, slight spines and some on the basis of colour like deep green and light green. However, there is not much variability in Okra in the Indian sub-continent and only a dozen varieties are listed in seed catalogues namely 'Lucknow Dwarf', 'Long Green', 'Red wonder', and 'Samalkota'. All of these give poor yield and are susceptible to the yellow-vein-mosaic virus, so they are not considered for the main season. The tow varieties released by the IARI Pusa, New Delhi in the fifties, namely 'Pusa Makhmali' and 'Pusa Sawani' are widely grown throughout India at present and has good fruit quality and colour too. But the 'Pusa Makhmali' is susceptible to the yellow-vein-mosaic virus and can be grown only for the spring-summer crop. As the variety has lost its resistance, the efforts of the Agricultural Universities and Research Institutes in the country since 1970's have made the way for the introduction of a number of varieties and lines resistance to the yellow vein-mosaic virus, they are 'Pb7' by the Punjab Agricultural University, 'Prabhani Kranti' by the Marathwada Krishi Vidyapeeth, Prabhani, Selection 4 and 10 evolved by the IIHR. Recently, the CCS Haryana Agricultural University, Hissar has developed a variety called, 'Varsha Uphar', which is resistant to yellow-vein mosaic. As a whole, the total area under hybrid Okra is only around 6 per cent of the total cropped area in India.

Table 4.9 State-wise Area, Output and Yield of Okra
(Area, Hectares, Output, M.T. : Yield M.T./ha.)

State	1993-94			1994-95			1995-96		
	Area	Output	Yield	Area	output	Yield	Area	Output	Yield
Andhra Pradesh	13,991	1,39,910	10.00	18,224	1,45,792	8.00	18,080	1,44,640	8.00
Arunachal Pradesh	1,129	3,880	3.44	1,129	3,880	3.44	1,129	3,880	3.44
Assam	10,290	1,00,845	9.80	10,550	1,00,200	9.50	11,000	1,10,000	10.00
Bihar	73,810	10,33,540	14.00	75,687	10,59,618	14.00	74,250	10,39,500	14.00
Gujarat	6,100	30,500	5.00	5,710	78,797	13.80	15,386	72,267	4.70
Haryana	4,650	50,500	10.86	4,900	55,000	14.22	6,450	72,900	11.30
Himachal Pradesh	357	4,450	12.46	380	4,740	12.47	390	4,875	12.50
Karnataka	12,613	1,10,364	8.75	13,664	1,19,560	8.75	14,368	1,25,720	8.75
Madhya Pradesh	11,888	68,300	6.00	10,568	84,000	7.95	11,096	88,000	7.93
Maharashtra	3,562	24,975	7.01	5,329	41,015	7.70	6,663	43,345	6.51
Orissa	78,900	6,83,900	8.57	79,900	6,87,000	8.60	82,297	7,21,350	8.77
Punjab	855	6,384	7.47	855	6,384	7.47	895	6,675	7.46
Rajasthan	4,393	17,280	3.93	3,717	10,271	2.76	3,474	10,757	2.87
Tamil Nadu	3,969	41,540	10.47	4,355	38,020	8.73	4,355	38,020	8.73
U.P. (Hills)	3,166	13,316	4.21	3,408	13,985	4.10	3,487	12,720	3.65
West Bengal	36,213	3,74,104	10.33	36,213	3,74,104	10.33	36,213	3,74,104	10.33
Delhi	302	1,771	5.86	1,880	10,720	5.70·	1,258	7,296	5.80
Pondicherry	356	3,408	9.57	404	4,040	10.00	393	3,930	10.00
Total	2,95,227	30,29,453	10.26	4,15,890	39,88,785	9.59	4,30,525	10,31,811	9.36

Source : National Horticultural Board.

Harvesting

Okra fruits are harvested on every second or third day from the time the first pod is formed. It takes 7-8 days from flowering to picking of fuits ready for the market. Normally, the market prefers small tender fruits ready for the market. Normally, the market prefers small tender fruits on every alternate day. So as to enhance the yield and for the developement of the plant, there is the need to harevest the Okra frequently.

Marketing

In the marketing of Okra, there are local traders middlemen, bulk purchasers, co-operative institutions etc. In the southern part of India economy, the bulk of Okra fruits are marketed weekly in the village or in weekly urban markets of in shandies. However, the scope for wastage is more in these markets since, the sales are taking place on the road-sides without any sheds. Again the absence of storage and godown facilities in these markets are considered to be the hurdles before the traders.

Exports

Okra accounts for 60 per cent of the export of fresh vegetables excluding Potato, Onion and Garlic. India exports Okra mainly to West Asia, Western Europe and the U.S. The demand for fresh Okra is more in the external markets.

Problems

(1) Problem of insect pests like Jassids, shoot and fruit-borer and cotton-boll worms.

(2) Diseases like yellow-vein mosaic virus and powdery mildew.

(3) Non-availability of quality seeds which are resistant virus and other diseases.

(4) *Problem of marketing* : As most of the growers are located in the rural areas and the consuming centres are located in distant places, there appears the problems like transportation, storage etc.

Among the vegetables exported, Okra constitutes nearly 60 per cent, and the scope for increasing this is vast, hence the above said problems have to be over-rided. In this regard, a proper long term strategy is needed in India. As the external demand for green, tender and six to nine cm long fresh okra is more, there is the need to develop these varieties. In this regard, Pusa A-4, Pusa Sawani, Prabhani Kranti, Panjab-7 and Sel-2 are ideally suited for export.

5

Cole Crops

A group of vegetable crops has originated from a common parent, the wild cabbage or colewort, *Brassica olercea* var. sylevestris. The wild ancenstor, the colewort, is a stout weedy perennial of the sea-coasts of the Great Britain and the South Western Europe. From this plant, there has arisen by selection or mutation, the great variety of cultivated forms. The Cole crops have spread all over the Europe, from the Mediterranean region. The commonest forms or commercially important crops of this group are cauliflower, cabbage, knol and to a lesseer extent Brussels sprouts and the heading Broccoli. Except some forms of cauliflower, all Cole crops require a cool climate to flower and set seed. India is the leading producer of cauliflower and cabbage in the world. Apart from this, Asian region dominates in terms of area as well as production of these.

These crops have become popular because of a good numnber of medicinal uses attributed to them. In general, the Cole crops are used against gouts, diarrhoea, colic trouble, stomach trouble, deafness and headache. The studies conducted on Cole vegetables in Japan and the USA have revealed certain protective proper against bowl cancer. The Cole crops are a rich the source of Vitamins A and C and they contain minerals – phosphorous, potassium, calcium, sodium and iron.

CABBAGE

An ancient and most important herbage vegetable is the Cabbage. It belongs to the family of Cruciferae. It is supposed to have been derived by cultivating the European wild colewort or wild Cabbage. Its cultivation is very old about 2500 BC. It was known to the Greeks and Romans. It is said that the ancient Germans, Saxons and Celts were the first to grow Cabbage in

North Europe and later it became an important herbage vegetable at a very early date in Scotland and Ireland. Today, Cabbage plants are grown the world over, except in the low tropics. In Inida, it was introduced by the Moghuls and popularised by the Brithishes. However, a possible assumption is that it might have been introduced by the Portuguese regime.

Values and Uses

Cabbage contains 92.1 g water, with some sugar and starch, considerable protein and valuable lime salts. It is rich in mineral matters and in Vitamins A, B^1, B^2 and C. Cabbage is used both for cooking and salad. It is eaten raw; steaming is preferable to boiling for the nutrients are returned. It is also used for feeding stock and chickens. In the European countries, Cabbage, in some form, is an important part of the daily diet of the poorer people. The Cabbage juice is said to be a remedy against poisonous mushrooms and used as gargle to remove hoarseness of throat. The leaves are used to cover wounds and ulcers and also recommended against hang-over. Researches on these in Japan and USA revealed certain protective properties against bowl cancer.

At present, China is the largest producer of Cabbage in the world and its production was 16.81 million tonnes in 1996. In terms of production China is followed by Russia, India, Korean Republic, Japan, Poland, USA, Germany, and others. The total production of Cabbage was 36 million tonnes in 1980 and at present it is about 51 million tonnes (Table 5.1)

Table 5.1 : Production of Cabbage in Major Producing Countries in the World

(in lakh M.T.)

Country	1995	1996	1997	1998	1999
China	141.49	162.14	169.05	169.05	169.05
India	42.00	42.00	42.00	42.00	42.00
Russia	52.80	50.35	30.29	28.26	28.26
Korea R.P.	31.50	30.25	28.95	29.92	29.92
Indonesia	0.76	0.88	13.35	11.56	11.56
U.S.A.	20.33	19.73	21.02	21.08	21.50
Poland	18.66	18.32	17.70	20.20	19.74
Ukraine	9.65	9.47	12.05	9.67	9.67
World total (including others)	480.60	494.95	477.08	476.32	475.74

Source : FAO.

Cabbage in India

The cultivation of Cabbage is undertaken in almost all states in India. Uttar Pradesh is the largest producer of Cabbage followed by Orissa, Bihar, West Bengal, Assam, Karnataka, Maharashtra, Madhya Pradesh and Tamil Nadu. The total area in Uttar Pradesh under this crop is above 41,000 hectares and the production is more than 10,70,000 metric tonnes. In recent years, the production of Cabbage in Assam and Bihar has declined while in the plains of Uttar Pradesh it has gone up as can be observed from Table 5.2. As a whole, the total area under Cabbage in India is about 220,000 hectares and the production is about 42,00,000 metric tonnes. The trend in area and production as estimated by the FAO shows a positive tendency is both respects over the years as can be seen from Table 5.3.

Cabbage is grown mainly as a winter crop in the plains of India. The main crop is cultivated in the Northern India where the winter temperature is relatively low. While in the hills, it is grown as spring and early summer crop. In states like Uttar Pradesh, the farmers are now cultivating more number of cropsi in a year due to availability of varieties that can be planted in different seasons. As Cabbage can be grown in all types of soil, still sandy loam is the best for an early crop and clay loam and clay soil for late crop.

Varieties

As Cabbage is biennial crop, there is a great variation in its type since there appear to be difference in its size, shape, colour of leaves and texture of head. The varieties commonly grown in India are two, viz., the early varieties and the late varieties.

The Early Varieties

(1) *Gold Acre* : It is selection from the 'Cophenhagen Market'. The heads of it are small, round and weighs 1-1.5 kg and it matures in 60-70 days after transplanting.

(2) *Pride of India* : It is slightly bigger than Golden acre and the head size is 1.5-2 kg.

Marketing of Vegetables in India

Table 5.2 : State-wise Area, Output and Yield of Cabbage

(Area : Hectares, Output, M.T., Yield, M.T./ha.)

State	1993-94			1994-95			1995-96		
	Area	Output	Yield	Area	output	Yield	Area	Output	Yield
Andhra Pradesh	1,180	11,800	10.00	1,549	38,725	25.00	1,284	32,100	25.00
Arunachal Pradesh	2,275	8,990	3.95	2,275	8,990	3.95	2,275	8,990	3.95
Assam	18,100	3,31,230	18.30	17,750	3,28,375	18.50	18,500	2,27,500	12.30
Bihar	70,915	11,34,640	16.00	40,915	6,54,640	16.00	39,415	6,26,320	15.89
Gujarat	6,000	1,20,000	20.00	5,215	66,347	12.72	7,349	99,608	13.55
Haryana	3,200	51,400	16.06	3,550	53,875	15.18	3,950	53,250	13.48
Himachal Pradesh	1,535	38,000	24.76	1,560	42,656	24.77	1,640	42,656	26.01
Karnataka	7,657	1,83,768	24.00	8,263	1,93,317	24.00	8,263	1,98,317	24.00
Madhya Pradesh	3,426	1,37,000	39.99	3,624	72,000	19.89	3,805	1,52,000	39.75
Maharashtra	11,113	1,88,985	17.01	9,933	2,14,351	21.58	8,624	1,61,202	18.69
Manipur	1,047	12,570	12.01	1,883	13,000	12.00	1,125	13,500	12.00
Meghalaya	1,060	11,766	11.10	1,100	12,486	11.35	1,122	13,460	12.00
Orissa	44,530	6,85,700	15.40	44,640	6,90,580	15.47	45,979	7,25,109	15.77
Rajasthan	4,472	23,280	5.21	6,317	25,864	4.09	6,067	17,384	2.87
Sikkim	506	4,545	9.00	225	2,200	9.78	225	2,350	10.44
Tamil Nadu	12,010	1,40,520	11.70	9,440	1,07,620	11.40	10,760	1,07,620	11.40
U.P. (Hills)	4,578	20,934	4.57	4,448	21,695	4.88	4,484	22,957	5.12
U.P. (Plains)	20,917	3,10,796	14.86	35,205	10,47,585	29.76	35,205	10,47,585	29.76
West Bengal	26,196	2,70,618	10.33	26,196	2,70,618	10.33	26,196	2,70,618	10.33
Delhi	510	15,484	30.36	799	23,782	29.76	782	23,040	29.80
Total	2,31,681	35,92,922	15.51	2,16,759	39,06,264	18.02	2,18,381	38,61,684	17.68

Source : National Horitultural Board.

Table 5.3 : Area and Production of Cabbage in India

Year	Area (in ha)	Production (in M.T.)
1990	167896	2373632
1991	178353	2796431
1992	215865	4357050
1993	230000	3590000
1994	220000	3910000
1995	220000	4200000
1996	220000	4200000
1997	230000	4200000
1998	230000	4200000
1999	230000	4200000

Source : FAO.

(3) *Sel 8* : This variety is characterised by a short stalk, medium sized leves, light green with wavy margins and slightly puckered leaf-blade. It is resistant to black-rot.

(4) *Copenhagen Market* : This is grown on a limited scale. The head size is 2-2.5 kg.

(5) *Express* : Some of the seed companies market this even though it is considered as Golden Acre as it resembles Golden Acre in all respects except it is said to be slightly earlier in maturity.

(6) *Sri Ganesh Gol* : Marketed by Mahyco, its plant type is similar to that of Golden Acre. However, it matures earlier.

(7) *Hari Rani* : This is popular in the Southern States, especially in the Nilgiri hills. It has a medium-sized frame with light green foliage.

(8) *Meenakshi* : It has good staying power.

(9) *Green Boy* : National Seeds Corportion markets this. It is an introduction from Japan and is early in habit, produces medium sized round heads.

(10) *Green Express* : National Seeds Corporation markets the seeds of it, before this, it was a popular variety in West Bengal. The head size depends upon planting time.

Late Varieties

These are not popular and grown for the late season or bulk supplies. The following are the varieties under this :

(1) *Pusa Drum Head* : The heads are flat, slightly loose weighing 3-5 kg. The outer leaves are light green with prominent midrib.

(2) *Septemeber* : It is an introduction from GDR and is a popular open pollinated variety in the Nilgiris. It has a large frame, bluish-green leaves with serrated margins. The head flatish-oblong, very right and weigh 4-6 kg. It is highly susceptible to black-rot.

(3) *Late Large Drum Head* : The heads of it are large flat somewhat loose and matures in 110-115 days.

As a whole, in India, of the total area under Cabbage, about 32 per cent of it is covered by hybrid varieties, which calls for an improvement in it.

Harvesting

The variety and transplanting determines the harvesting time. However, for early varieties, it varies in between 70 to 80 days while for late varieties, it is 100-120 days. As a whole, it is harvested when the heads are large and firm. After harvesting, the grading is done according to its size and quality.

Marketing

In the marketing of cabbage, normally there appear two channels namely (1) Producer–Commission Agents (2) Producer–Pre-harvest Contractors. In most of the cultivable areas, the share of the commission agents in the total sales from the producers is above 80 per cent and the remaining is handled by the pre-harvest contractors. As a whole, there is no organised marketing system for this crop. As far as the cost incurred by the cultivators for its marketing is concerned, a major portion goes for commission followed by transport, loading packing etc.

Problems

(1) The productivity of Cabbage in India is much lower, which is around 19 tonnes per hectare.

(2) Cabbage is susceptible to some of the fungal diseases like black leg, yellows, damping off, soft rot, downy mildew etc., Insects Pests like Cabbage Maggot, green cabbage worm and cabbage looper and Cabbage and Turnip Aphids. All of these reduce the productivity and production.

(3) Problems of marketing.

(4) Increasing level of marketing cost and the involvement of intermediaries minimised the scope for an improvement in this crop.

(5) Absence of proper timely transport and storage facilities.

Even though India has the ability to grow this poor man's popular vegetable in large volume, the above said problems are actually hindering the scope for an improvement in it. So, there is the need to overcome this. Then only the country can increase the production and can export it to the international market. Again, there is vast scope for introducing F1 varieties to a larger extent since the present area under this is ony one third of the total. Hence, the country has to utilise the available opportunities to a greater extent.

CAULIFLOWER

Cauliflower is a favourite vegetable in all temperate regions. The name cauliflower has originated from the Latin words 'Caulis' and 'Floris' which mean cabbage and flower respectively. It has originated from wild cabbage known as Coleworts. Cauliflower has been developed from these wild plants. In this, a short erect stem is produced with an undeveloped inflorescence. The whole inflorescence forms a large head of abortive flowers on thick hypertropical brancehes. The leaves are often tied around the mass of flowers to keep them white. The successful cultivation of it requires greater care since it is a delicate crop.

Values and Uses

The nutritive value of cauliflower per 100 g of edible portion is 91.7g water, 0.2g total fat, 2.4 g protein, 4.9 g carbohydrates. Cauliflower is rich in Vitamin A and Vitamin C. Cooked cauliflower contains a good amount of Vitamin B and a fair amount of protein in comparison to other vegetables. Cauliflower is used as vegetable in curies, soups and for pickling purposes. The dried cauliflower can be preserved for use in off-season.

In the world, cauliflower is gown in an area of 700 thousand hectares and its production is about 13428 thousand tonnes. Of the total area under this in the world, more than 70 per cent is in Asian region. India possesses largest area unde this crop followed by China, France, Italy, U.K., U.S.A. and others. In terms of production too, India stands first followed by China and others. (Table 5.4)

Table 5.4 : Production of Cauliflower in Major Producing Countries in the World

(in Lakh M. Tonnes)

Country	1995	1996	1997	1998	1999
India	50.00	50.00	50.00	50.00	50.00
China	36.58	43.58	44.72	44.72	44.72
France	5.33	5.23	5.09	4.69	4.70
Italy	4.71	4.76	4.94	5.28	5.28
U.S.A.	2.96	3.01	2.87	2.60	2.60
U.K.	3.94	3.94	2.19	1.92	1.92
Spain	2.85	3.07	3.13	3.52	3.60
World total (including others)	126.93	134.28	136.25	136.86	136.90

Source : FAO.

Area, Production and Yield of Cauliflower in India

Cauliflower is a popular winter vegetables grown in Inida. It was introduced in India in 1822. Now, it is cultivated mostly in the North. As far as the production of it is concerned, Bihar is the largest producer followed by Orissa, West Bengal, the plains of Uttar Pradesh, Haryana, Madhya Pradesh, Assam and Gujarat. The total area under cauliflower in Bihar is 72,750 hectares and the production is 11,63,360 metric tonnes, while in Orissa the production is 6,17,040 metric tonnes from an are of 49,409 hectares

Table 5.5 : State-wise Area, Output and Yield of Cauliflower

(Area : Hectares, Output, M.T., Yield, M.T./ha.)

State	1993-94			1994-95			1995-96		
	Area	Output	Yield	Area	output	Yield	Area	Output	Yield
Arunachal Pradesh	990	3,590	3.63	990	3,590	3.63	990	3,590	3.63
Assam	12,725	1,61,607	12.70	10,000	1,60,000	16.00	12,850	1,20,000	9.34
Bihar	38,853	6,21,698	16.00	72,750	11,63,360	16.00	72,750	11,63,360	16.00
Delhi	4,849	1,39,269	28.72	2,066	26,685	12.92	1,602	26,826	13.00
Gujarat	4,600	1,15,000	25.00	2,714	33,753	12.44	6,416	1,02,765	16.02
Haryana	5,300	1,06,000	20.00	6,250	1,28,125	20.50	6,400	1,34,400	21.00
Himachal Pradesh	645	12,692	19.68	670	13,195	19.68	690	13,800	20.00
Jammu & Kashmir	1,139	19,692	17.29	1,139	19,692	17.29	1,139	19,692	17.29
Karnataka	3,282	78,768	24.00	4,208	78,768	18.72	4,208	78,768	18.72
Madhya Pradesh	8,074	3,22,000	39.88	8,379	1,25,000	14.92	8,798	1,31,000	14.89
Maharashtra	6,985	1,36,609	19.56	6,426	1,26,490	19.68	3,583	71,855	20.05
Manipur	820	8,200	10.00	840	8,400	10.00	890	8,200	10.00
Meghalaya	875	10,894	12.45	955	11,343	11.88	1,000	13,000	13.00
Orissa	47,900	5,85,300	12.22	47,970	5,87,130	12.25	49,409	6,17,040	11.57
Punjab	2,160	57,971	24.56	2,160	57,971	24.56	2,310	56,879	24.62
Rajasthan	1,118	5,819	5.20	375	1,058	2.82	154	601	3.90
U.P. (Hills)	3,255	13,218	4.06	3,346	13,720	4.10	3,345	15.094	4.51
U.P. (Plains)	18,357	2,07,199	11.29	18,357	2,07,199	11.29	18,357	2,07,199	11.29
West Bengal	18,357	2,70,600	10.33	18,357	2,70,600	10.33	18,357	2,70,600	10.33
Total	1,88,545	28,72,942	15.24	2,16,142	30,33,910	14.04	2,20,025	24,73,987	11.24

Source : National Horticultural Board.

as can be observed from Table 5.5 where data are given for state-wise area, production and productivity of cauliflower since 1993-94 up to 1995-96.

The trend for area and production of cauliflower since 1990 up to 1998 as per FAO estimate shows that the production has been increasing over these years as shown in Table 5.6.

Table 5.6 : Area and Production of Cauliflower in India

Year	Area (in ha.)	Production (in M.T.)
1990	238632	3394897
1991	202787	2998061
1992	222436	4220053
1993	240000	4500000
1994	280000	4800000
1995	280000	5000000
1996	280000	5000000
1997	290000	5000000
1998	290000	5000000
1999	290000	5000000

Source : FAO.

Varieties

The initial introdiction was 'Cornish' types from England followed by the European types. the Inddian cauliflower or the tropical types are the result of intercrossing among these types.

In India, two separate groups of cauliflowers are normally grown viz., Indian or tropical types, originated in India during the past 175 years and the annual temperate types known as 'Erfurt' or 'Snowbali' types. The major differences between these two groups are their adaptability to different temperatures for curd formation. The tropical types are resistant to water-logging and heat. Almost all of the cultivars in India are known by the name of Hindi months. They are :

(1) September Maturity

In this, 'Early Kunwari' is the earliest cultivar which has bluish-green leaves with waxy blooms. The curds are white and not so compact and is sensitive to riceyness.

(2) October-November Maturity

It consists of two varieties namely Pusa Katiki and Pusa Deepali. The first is one of the early releases from the IARI, having medium-sized plant and bluish-green leaves with medium sized curds. While the latter is developed through in-breeding from the local collection by the IARI and it has uniform medium, long, erect foliage with rounded tips. The curds are white and deep.

(3) November-December Maturity

There are three varieties under this viz., 'Improved Japanese', 'Pant Shubra', and 'Pusa Himjyoti'. Improved Japanese is an introduction from Israel with long, erect, narrow leaves and medium sized curds. Pant Shubra was developed at Pantnagar. Pusa Himjoti is product of selection at the IARI regional station, Katrain. This has erect bluish-green leaves witha waxy coating. The heads are pure white and retain their colour even after exposure.

(4) December-January Maturity

(a) Pusa synthetic, a variety develped by IARI, has erect leaves varying in colour, frame, narrow, medium curds, somewhat creamy white to white compact.

(b) Pusa Shubra is another variety develped by the IARI. The plants are errect with somewhat long stalk and light bluish-green leaves. In this group, two more varieties have been released by the PAU, Ludhiana viz., Punjab Giant 26 and Punjab Giant 35.

(5) January-February Maturity

The varieties grown in this group are known as the 'Snowball' types. They are :

(1) *Dania* which is developed at Kalimpong, and is sucessful in the eastern region, especially in the hills. It has very sturdy plants with wavy leaves and medium-deep curds. However, it is sensitive to fluctuating environment.

(2) *Pusa Snowball 1*. This is made at Katrain. Its of straight, upright leaves, tightly covering the curd.

(3) *Pusa Snowball K-1.* This is field resistant to black-rot. It has upright, wavy, light green leaves, self-blanched, snow-white, hidome curds. The curds quality and late maturity has made it as a popular variety among the growers.

Apart from these varieties, several others are also marketed by the private seeds companies under different names. As whole, the early varieties are smaller in size and have a yellowish tinge and the mid-season varieties produce the largest curds but are mostly dirty white in colour. While the late varieties are compact and milky-white. In Inida, of the total area under cauliflower, the area under hybrids is only around 4 per cent.

Harvesting

The harvesting of cauliflower takes place when the curd reaches a proper size with an attractive colour. Normally harvesting is done as and when the curds develop. So as to overcome the problem of damage while harvesting this, care is taken to harvest it either in the evening or early in the morning. The harvesting period depends upon the transplanting; usually early varieties are harvested in beteween 60-70 days while mid season varieties in between 80-90 days and the late varieties in 120-140 days.

Marketing

For the marketing cauliflower is graded into three viz., depending on the size, colour, quality and variety. For the purpose of packing, the growers as well as traders use big nets and transport the product through trucks.

In its marketing, there appears to be two important channels. In the first, the growers sell it to the commission agents which constitutes a major portion and in the second, the growers sell the whole crop prior to harvest to the pre-harvest contractors. However, direct sales to retailers is negligible. As a whole, the commission agents or the intermediaries play a dominant role in the marketing of cauliflower in India. As far as the sale of this is concerned, it is sold by numbers through auction by these commission agents.

Exports

India exports cauliflower to Bangladesh, Germany, Netherlands, UAE and Nepal. In 1993-94 about 4 tonnes of cauliflower and headed broccoli exported to Bangladesh and in 1994-95, 24.50 tonnes valued at Rs. 1.48 lakhs exported to Germany. In 1996-97, 11.20 tonnes to Netherlands and in 1997-98, 26 tonnes valued at Rs. 6.12 lakhs exported to UAE and 23.10 tonnes worth of Rs. 0.38 lakhs to Nepal. Apart from these, in 1995-96, India exported 3,660 kg. seeds of cauliflower to Bangladesh and Pakistan while during 1996-97 it was 11,800 kg valued at Rs. 1.90 lakhs and the destination was Netherlands and Malaysia.

Problems

(1) The productivity of cauliflower in India is much lower as compared to other producing countries in the world. This is because of (a) lack of seeds of improved varieties, (b) lack of awareness among the growers about the latest varieties and (c) lack of agro-techniques.

(2) Non-availability of proper education and training for the farmers about pre- and post-harvest aspects.

(3) Major diseases like damping off, black rot, soft rot, clubroot, downy mildew, sclerotinia rot and Insect pests like cabbage magyot, Green cabbage worm etc. reduced the scope for an expansion in production.

(4) Poor marketing facilities minimised the scope for obtaining a remunerative price to the growers.

(5) Absence of proper timely transport facilities and non-availability of storage facilities.

Prospects

Cauliflower is considered as one of the valuable vegetable crops of India and it contains good amount of vitamins and protein and the scope for external marketing is bright. So, there appears the need to overcome the prevailing problems of this sector. In this regard, the following solutions will be useful. They are :

(1) In the development of cauliflower, the major thrust should be on developing lines which can form curds even

during summer, at a temperature around 25°-28° C, exploiting heterosis through the use of self-incompatible lines in September to November maturity groups; developing lines with high curds ratio; and breeding varieties resistant to black-rot, downy-mildew, Alternaria blight and Sclerotinia rot.

(2)   There is the need to produce quality seeds. In this regard, there is the need for a very thorough roguing of off-type and undesirable plants are essential for quality seed production.

(3)   Proper irrgational facilities have to be provided since it requires very heavy irrigation.

(4)   There is the need to develop the marketing facilities and it is possible to overcome the problem with the involvement of intermediaries.

(5)   By having the cooperatives, it is possible to develop the storage facilities. Apart from this, the coorperatives can also provide mobile transport facilities which ultimately solves the problem of transportation.

(6)   The horticultural departments should provide timely education and training to the growers about the basic aspects like preparation of the soil, sowing of seed, manuring, harvesting and post-harvesting.

(7)   Efforts are needed to expand the productivity and production in the growing regions, since it can be grown only in the cool climate and the improvement work on this is restricted to the temperate regions of the country.

KNOL KOHL

Kohirabi or Knol Kohl originated in the coastal countries of Northern Europe. In Kohirabi no head is formed, but the short stem is transformed into juicy mass of edible tissue, which stands out of the gournd. It is large, spherical and turniplike, white or purple in colour with large leaves. It is an early spring or fall crop, as it does not like the heat of summer.

Knol Kohl contains rich protein, carbohydrates, fibre and Vitamin A, Vitamin C and Vitamin K etc. It is used as a vegetable and its value is more in its early stages. In USA, it is also used for stock feed.

This vegetable is popular in Kashmir, West Bengal and Karnataka. The popular varieties are 'White Vienna', Purple Vienna, and 'King of North'. the white Vienna is an early variety having dwarf plants, medium green leaves and stem, knobs smooth, light green, flesh creamy white, tender with delicate flavour. The Purple Vienna are globular round, large in size with bluish purple tinge, flesh light green. While king of North is dark green foliage with broad leaf blade. Knobs are flatish round and dark green in colour.

All of these varieties are foreign introductions and no improvement work has taken place in India because of the limited importance of this vegetable.

BRUSSEL SPROUTS

This is a recent develpment in cole crops. The crop got its name from the city of Brussels in Belgium. It has been grown there for several hundred years. This crop resembles miniature cabbages. It grows well in cool hill regions. Brussels sprouts has more nutritious value since per 100g for edible portion contains 85.5g water, 4.7g protein 7.1g carbohydrates, 1.2 g fibre and vast amount of Vitamin A and Vitamin C.

In recent years it has become popular around big cities to meet the ever increasing demand from big hotels for the tourists. It is a delicacy with the foreigners and is slowly gaining ground with the vegetable growers nearer to the big cities. Some times, this is called as 'mini cabbage' since it looks like cabbage head.

The Hilds Ideal variety has been identified as a suitable variety for our country. The sprouts of this are compact and possess good flavour. A small portion of loose sprouts at the base of the plant if marketable. As far as harvesting of it is concerned, it takes about 120 days to form sprouts in the Northern part of India. As a whole, the importance of this cole vegetable has yet to be realised in India. Hence, the scope for area expansion and an increase in its productivity and production is always there. In this regard, efforts are needed to popularise this crop and to note down the prevailing status of it as well as future prospects too.

Other Cole Crops

Apart from these above-said cole crops, the other two types grown are : (1) Sprouting Broccoli and (2) Kale. In the former, there are two important varieties viz., Green sprouting broccoli and Purpule sprouting broccoli. The main difference between the two varieties is the colour of the bud. The cultivation of this cole crop in India is very limited.

As far as Kale is concerned, it is popular in Kashmir valley. It is grown in every kitchen garden and is the main contaminant of cabbage seed. It is a winter vegetable and is available at a time when other vegetables are scarce. As it is hardy, it can withstand even unfavourable weather.

Future for Cole crops

Even though there appears to be the prevalence of several problem in the pre- and post-harvesting aspects of these crops, still there is vast scope for an improvement in terms of area, productivity and production of these in India. As the demand for cabbage, cauliflower another cole vegetable are ever increasing in India and cabbage is considered as a poor man's vegetable in the country, the scope for these is bright. Again, the importance of Knol Kohl, Brussels sprouts, Sprouting Broccoli and Kale are yet to be recognised in this country and scope for an improvement in these are also prevailing, all of these definitely indicate a bright future for this sector. So as to encash this, there is the need for a proper planned strategy.

6
Cucurbits

Cucurbits are a group of tropical vegetables with a long taproot system. This group comprises *salad* kinds like cucumber and long melon and the cooking types bottlegourd, bittergourd ashgourd and pumpkin and dessert fruits, muskmelon and watermelon and number of other crops mostly of trailing habit. They all belong to the family of Cucurbitaceae and are grown during summer.

These vegetables are extensively grown in river-beds in different diara regions, river-banks or riverine pockets. They are also cultivated in the gardens and fitted into crop rotations, especially in the gardens and fitted into crop rotations, especially in the Kharif season. Cucurbits are grown in summer and the rainy season in most parts of Inda and even in winter in some parts of southern and western India. These crops are viny in nature and occupy large areas and hence the system of planting varies region-wise. These are used as vegetables as well in the preparation of ketchups. Some of them are used as a substitute for bath-sponges and they are also having many medicinal values too.

There is an extensive demand for these cucurbits both in the internal as well as the external markets. These are ntritious and are commercially significant in the whole world. China is the leading producer of some of these cucurbits and India is also one of the major producers of these crops in the world.

CUCUMBER AND GHERKIN

Cucumber belongs to the genus cucumis and species sativus. It is an important gourd fruit and is indigenous to South India. It has been cultivated for more than 4000 years. The earliest writings of the Hebrews, Egyptians, Greeks and Romans have numerous references to this plant. It had reached Europe by the 17th century and is now widespread. Cucumber is a rough stemmed trailing vine with yellow axillary flowers and round to elongate prickly fruits.

Gherkins belong to the family of cucurbitacae and sp *Cucumis anguria* commonly cultivated in West Indies known to be West Indian gherkin. Gherkin has tiny fruits with a thin flesh and numerous seeds. They are also called as burr cucumber.

Values and Uses

The water content in cucumber is as high as 96 per cent, while it is 93 per cent in Gherkins. The protien content in gherkins is 1.40 per cent while it is 0.5-0.7 in cucumber. Both of these are vitamin rich since vitamin B1, Vitamin B2, Vitamin C and Vitamin A are all there in these vegetables.

Cucumber is an important vegetable, the tender fruits are eaten raw or with salt in *salad*. They are also used as cooked vegetable. The oil from its seeds are used in the preparation of ayurvedic medicines.

Gherkins are used for pickling purposes. The demand for gherkins is very high in the western countries. In the west, there cannot be humburger without gherkin in it. U.S.A. is the largest consumer of this. Gherkin is preserved in natural vinegar and acetic acid. It is also packed fresh in glass jars mixed with spices, onion, mustard and dil leaves. The processed products are consumed as snacks with beverages, between sandwitches hamburger, pizzas etc.

China is the largest producer of cucumbers and gherkins in the world. The total production of it is more than 50 per cent of the total world production. In terms of product, Asian region dominates it. The total production of these in 1995 was 232.74 lakh tonnes and it went up to 258.7 lakh tonnes in 1997. In 1998 it was 266.62 lakh tonnes. Table 6.1 gives data on the production of cucumber and gherkins in the major producing countries over the years.

Cucurbits   107

Table 6.1 : Production of Cucumbers and Gherkins in Major Producing
Countries in the world

(in '000 Tonnes)

Country	1997	1998	1999
China	14262	14262	14262
Iran	1038	1100	1100
Turkey	1400	1400	1400
U.S.A.	1087	1077	1077
Japan	780	800	800
India	116	116	116
World total (including others)	26521	26641	26662

Area and Production in India

Cucumber and gherkins are commonly grown throughout India especially during summer. The total area under these in India at present is about 17200 hectares which was 15900 hectares in 1990. In terms of production, at present it is about 115000 metric tonnes and it has been increasing over the years as can be observed from Table 6.2. As far as gherkin's cultivation concerned, it is there in about 6,000 hectares. The cultivation of it is mainy undertaken in Karnataka, Tamil Nadu and Andhra Pradesh. The production of this during 1998-99 was about 40,000 tonnes.

Table 6.2 : Area and Production of Cucumbers and Gherkins
in India 1990-1999

Year	Area (in ha.)	Production (in M.T.)
1990	15900	102000
1991	16288	105690
1992	16500	109000
1993	16700	111000
1994	16900	113000
1995	17000	114000
1996	17200	115000
1997	17000	116000
1998	17000	116000
1999	17000	116000

Source : FAO.

Varieties

Japanese long green, straight, painsetle, Balam Khira, Poona Khira, Priya, Khirapati, Fihybrid etc. are some of the important varieties of cucmebr grown in India. In the northern part of India "Painsetle" a variety introduced from the U.S.A. which produces cylindrical dark green fruits with very few white spines are popular. It carries high resistance to downy-mildew, powdery-mildew, anthracnose and angular leaf-spot. In the western parts of Maharashtra, Poona Khira and Balam Khira are popular. These produce small pale green fruits with brown spines and are also grown in Uttar Pradesh. Sheetal is a variety which is suitable for the Coastal regions. Darjeeling or Sikkim types are grown in the hills of North Bengal. In the mid-hills of Himachal Pradesh, Khira 75 and Khira 90 are grown. In the cooler regions, Japanese Long Green and Straight 8 are cultivated. The main varieties of Gherkins are Calypso and Venlo pickling and NCVH-32, NCVH-35, NCVH-40 and NCVH-41 which are high yeilding varieties and are the most popular in India.

Harvesting

In the marketing of cucumber and gherkins, there appears to be the prevalence of intermediaries. Normally the share of these intermediaries are above half in the consumer price. As a whole, there is no organised marketing system for these crops.

Exports

Gherkin is the main item of export from India. India exports fresh and chilled gherkins, provisionally preserved gherkins, preserved by Vinegar gherkins etc. The total value of all these products' export during 1997-98 was Rs. 6,810.70 lakhs. As far as the exports of fresh or chilled gherkins are concerned, the main markets are Spain, Belgium, Netherlands, U.S.A., France, Russia and others. For preserved gherkins, Spain, U.S.A., Belgium, France and for preserved by Vinegar/Acetic acid. Spain, France, U.S.A., Belgium are the major markets. Over these years, the volume as well as the value of these exports has been increasing as can be observed from Table 6.3.

Table 6.3 : Export of Cucumbers and Gherkins
(Values in Rs. Lakhs)

Item	1992-93	1993-94	1994-95	1995-96	1996-97	1997-98
Gherkins, Fresh or chilled	3.59	17.62	178.05	645.55	1,446.93	1,783.76
Gherkins, Provisionally Preserved	121.57	295.85	1,178.26	2,038.11	2,690.19	3,179.27
Gherkins, prepared/ Preserved by Vinegar/Acetic Acid	-	24.40	687.47	367.21	655.99	1,847.67
Total	125.16	337.87	2,043.78	3,055.87	4,795.11	6810.70

Source : DIGCIS, Ministry of Commerce.

Problems

(1) The production of cucumber and gherkins are lower because of a number of diseases. They limit the scope for area expansion. The important diseases are bacterial wild, anthracnose, downy mildew, powdery mildew, angular leaf spot and mosaic. Apart form these, many insects damage the plant which leads to a number of diseases. The important insects are red pumpkin beetles, aphids, cutworms and fruit fly. Non-availability of timely measures actually reduces the socpe for an increase in productivity and production of these crops.

(2) Absence of proper storage facilities restricts the scope for obtaining remunerative price.

(3) Non-availability of proper marketing facilities.

(4) Absence of incentives to the farmer from the side of the Government and organisations to expand the area under gherkins and to increase exportable volume of these.

As the prevailing efforts to enlarge the area under these crops are insufficient, there is the need to provide the required assistance to the growers. Again, there is the need to overcome the prevailing problems with timely measures. The demand for gherkins in the international market is ever increasing and the scope for increasing our exports is alwasy there since the hamburger culture is growing in U.S.A., France, Spain, U.K., Holland and Belgium. The consumption of gherkins in the U.S.A is above 750,000 tonnes per annum and the demand in Australia and South American countries is also moving positively, so, what is required now is a proper plan to improve this sector and to increase our exports. Side by side, there is also the need to increase the domestic consumption.

PUMPKIN, SQUASHES AND GOURDS

Pumpkin and squashes are gourd fruits belonging to the genus Cucurbita. There has been considerable speculation as to whether they are natives of America or Africa. However, the available evidence indicates that they are of American origin. It is a warm-season fruit. Pumpkin has the unique advantage of good storage capacity and most of the fruits are big sized. The plants are coarse annual vines with large yellow flowers and fruits that rest on the ground.

Values and Uses

The nutritive value of pumpkin in 100 g of edible portion contain 92.6 g moisture, 25g calories. Vitamin A, Vitamin C, Fat, Fibre, Iron, Sulphur etc.

It is used as a vegetable. The immature fruits are used as a fresh vegetabe, stewed, boiled, or fried, while matured fruits are backed, canned, or fed to livestock. The young leaves, tender stems and flowers are also cooked. The flowers are used in the preparation of "chutney" in South Indian houses. The seeds are rich in fats and proteins and can be utilised as the source of an edible vegetable oil. Pumpkin seeds, in deep fat and salted are popular in the Western markets. the field pumpkins are used for pies, canning, and cattle feed.

In the production of pumpkins, squash and gourds, the Asian region dominates, the highest producers of these in this region are China, Turkey and India. The total production of these in this region during 1997 was 8908 thousand metric tonnes while the world production during the period was 13785 thousand metric tonnes. In the production of these, India stands first in the world. In the African region, the major producers are Libya, Sudan and

Moracco while in South America, Argentina and Peru are the major producers of these. Table 6.4 shows the trend in the production of these in different regions of the world.

Table 6.4 : Production of Pumpkins, Squash and Gourds in Different Regions of the World

('000 M.T.)

Region	1995	1996	1997	1998	1999
Africa	1291	1378	1374	1420	1440
N.C. America	480	568	598	592	592
S. America	840	831	909	862	864
Asia	8132	8783	9235	9276	9271
Europe	2162	2063	2340	2243	2236
World	13090	13798	14713	14647	14664

Source : FAO.

Area and Production in India

Pumpkin, squashes and gourds are grown in almost all states in our coutnry. The total area under these at present is about 3,35,000 hectares and the production is around 3200000 metric tonnes. As these can be stored for a longer period, the cultivation of these are widespread through India. Table 6.5 gives data on area and production of these over the years in India which shows an upward movement both in terms of area and production.

Table 6.5 : Area and Production of Pumpkins, Squash and Gourds in India

Year	Area (in ha.)	Production (in M.T.)
1990	306000	2860000
1991	308000	2933000
1992	310000	2950000
1993	315000	3100000
1994	320000	3100000
1995	330000	3200000
1996	335000	3200000
1997	345000	3300000
1998	345000	3300000
1999	345000	3300000

Source : FAO.

Varieties

A large number of varieties are grown in India. According to shape, size, colour of flesh, they are classified in India. The important varieties are Large Red, Large round, Yellow Flesh and Red flesh. In the south Co 1 and Arka Chandan are popular. Some of the imported varieties like early yellow prolific, Australian Green Butternut, Green and Golden Hubbard are also cultivated. Pusa Alankar and F1 hybrid is also becoming popular now-a-days. Among these, Arka Chandan gives medium sized fruits, while fruits of Co 1 are bigger and globular in shape and late maturing. Co 2 is early maturing and bears small fruits. Pusa Vishwas produces medium sized fruits, the flesh colour is a golden yellow and the fruits are long lasting. Apart from these, CM 14, a variety from Kerala Agricultural University produces Flattish, round fruits medium in size with shallow furrows and the fruits are weighing moderately in between 6-7 kg. This is a suitable variety for South India for the January-March and September-December crop.

Harvesting

Harvesting of these are done according to the taste and preference of the market. They are harvested either in green form or mature stage. As the fruits can be stored for a longer period, these are having more value especially in the rural areas.

Marketing

As far as the marketing of these are concerned, there is no well organised marketing system. As the weight of these fruits are more, the marketing of these is considered to be a difficult task for the growers. Because of this, exploitation is taking place from the side of intermediaries in its marketing and the share of the producers in consumers price is normally lower.

BITTERGOURD

This is a popular vegetable grown in different parts of India. The bitter taste of this is liked by many. It has some medicinal values as it is used by the diabetic patients so as to control it. Bittergourd is rich in iron and vitamins. The bittergourd finds its won reputation in various dishes and in the preparation of curries. The villagers in South India dry it along with salt and use it for a long period. The major producers of bittergourd in the world ar China, Malaysia, India and tropical African countries. In West Indies, tropical Mexico and South America, it is one of the most commonly used medicinal herbs. In U.S.A., U.K. and France, it is grown as an ornamental annual. The fruits of it are a tonic, stomachic, carminative and produce a cooling effect. The fruits, leaves and roots have long been used in our country as folk medicine for diabetes.

It can be grown in all types of soil. Several varieties are available for cultivatin in India. The old variety "Coimbatore Long" which is whitish green in colour is preferred by the South Indians and not by the North Indians. The other varieties are VK 1, Priya, both of these are suitable for South India. The fruits are extra long, round and weigh more than 200g. Apart from these "Pusa Do Muasmi" and "Pusa Vishesh" are also grow in India. The latter is dwarf - vine variety, its fruits are a glossy green, medicine-long and thick, suitable as vegetable and for prickling and dehydration. The recently introduced varieties are "Punjab BG 14", "NDBI", "Phule BG6", "Kaliyanpur Sence", etc.

Marketing

The markets for this is very limited. Marketing of this is normally done by the producers themselves, however, the involvement of middlemen is common. The weekly markets and the daily local markets are the main marketing centres for bittergourd.

Problems

(1) The diseases as well as insect pests which are there in the cucurbits also damage the bittergourd crop. However, the major problem comes from the fruit fly which actually reduces the production.

(2) There is vast scope for a better utilisation of bittergourd in the preparation of pickles and wafers, which has not yet been fully utilised in India.

(3) The prevailing marketing system has minimised the scope for an area expansion. As bittergourd is one of the most nutritive and commercially important cucurbit vegetable, hence efforts are needed to improve this in India.

BOTTLE GOURD

It has originated from Arfica and later it spread to other warm countries. It is an important vegetable grown in India. Bottle gourd is a summer crop but it cannot withstand post. It is large vine with white flowers and is cultivated throughout India.

The fruits of this are used as a vegetable and in the preparation of sweets too. The hard shell of this is used as water jugs, domestic utensils, in the production of musical instruments, floats for fishing nets etc., the pulp, stem and leaves are used in the preparation of local as well as other medicines.

Varieties

In bottlegound, there are both old and new varieties. The recent introduction are : Pusa summer prolific long", "Pusa summer prolific Round", "Arka Bahar". The fruits of these are long and cylindrical and weigh about one kilogram. "Pusa Naveen" is suitable for planting in spring-summer and in the rainy season in the northern plains. Punjab Kamal, NDBG1 and KBG 13, are some of the other varieties.

Problems

(1) Problem of pests and diseases as that of cucumber reduces the scope for an increase in its productivity and production.

(2) Non-availability of storage and marketing facilities.

(3) Absence of research on alternative uses of these is in increasing manner.

SNAKE GOURD

It is a trailing herb cultivated throughout India. It is a popular vegetable and has a pride place in the South Indian dishes. The Fruit has more nutritive value.

It grows well in loam soil and can be grown in other soil too. In India snake gourd is generally cultivated during the rainy season.

Varieties

There are two types of Snake gourd grown in our country namely : (1) Light green with white stripes and (2) Dark green with yellowish or pale green stripes. The important varieties grown in India are as follows : Co 1, is an early-maturing cultivar, which bears long fruits, dark green with white stripes and flesh whitish-green. "TA 19" is another variety, it is light green with white stripes at the stylar end.

POINTED GOURD

It is said to have originated in India. It grows well in warm and humid climate. It can be grown well where rains are more.

The vitamin rich pointed gourd contains major nutrient and have elements like 9mg magnesium, 2.6mg sodium, 83mg potassium, 1.1mg copper, 17 mg sulphur per 100g edible part. It plays an important role in human physiology. Pointed gourd is easily digestible, diuretic and laxative, invigorates the heart and brain and finds useful in disorders of the circulatory system.

It is extensively cultivated in eastern Uttar Pradesh, Bihar, West Bengal and to some extent in Assam, Orissa, Madhya Pradesh, Maharashtra and Gujarat. In the diara lands of Bihar and certain parts of Uttar Pradesh, the crop is rainfed. The plants survives for a long period, so it is a remunerative crop.

Depending upon the size, shape and striation of the fruits, the pointed gourd plants are classfied into four categories viz., (1) Those with dark green fruits with white stripes and are 10-13 cm long, (2) 10-16cm long, thick dark green fruits with very fiant pale green stripes, (3) small 5-8cm long, roundish, dark green, striped fruits, and (4) fruits are small, tapering towards end, green and striped.

Varieties

The popular local varieties are 'Guli', 'Dandali', 'Kalyani',

'Bihar Sarif', 'Sopari Safeda', 'Niria' etc. FP-1, FP-3, FP-4 are the recent varieties developed for commercial cultivation.

Harvesting

Harvesting of the fruits is done at mature green marketable stage. The yield is more after the first harvest.

As the crop can be rasied through inter cropping system, this will increase the income of the growers. So, the scope for increasing the production is more in Inida. Apart form this, the plants can survive for several years so that the cost involved in its production is minimum, hence efforts are needed to popularise this crop in India.

RIDGED GOURD

The genus Luffa is from old world and includes two cultivated species viz., ridgegourd and sponge gourd. The probable centre of origin and the primary gene centre of this is tropical India. It is an important tropical vegetable and is an annual. It is an important crop in India and in South-East Asia. The plant is characterised by long, training, viny stems.

The average nutritional value of this is 2.04. It is popular in the south and east India. The edible part is tender fruit, cooked as vegetable and also prepared as *chutney*. It is easily digestible.

Varieties

Several varieties are available for cultivation. The important improved cultivars are : (1) 'Pusa Nasdar', this is an early-maturing, produces club-shaped fruits, 15-20 on a vine, light green in colour (2) 'Co 1' a light green and attractive variety. It yields around 15 tonnes per hectare (3) 'Co 2' : The fruits of this are long, slender, green and less-seeded. The yield is about 25.5 tonnes/ha (4) 'PKM-1'. The fruit are dark green with shallow ridges, the yield is 37 tonnes/ha.

Prospects

As the cultivation of this is easy and has wide adaptability, there is the need to improve the production of this. In this regard, a concerted effort for the collection of germplasm though intensive surveys and its introduction is needed. Measures should be taken to collect and conserve the local varieties. Again, the development of high-yielding varieties and hybrids though breeding should be stimulated. The characters in crop improvement should include high yielding ability, good keeping quality and high nutritive value.

MUSKEMELON

It is a native of Central Africa, was introduced into India probably during the Mughal invasion around the 14th century from the Central Asian region which comprises the present Russia, Iran, Afghanistan and parts of Pakistan. Since, then it has spread to different parts of the country.

Muskemelon is supposed to be very wholesome and nutritious. It prevents constipation. The nutrition value varies from variety to variety.

Muskmelon requires high temperatures and dry climatic conditions. It grows better only in summer. It is mainly grown on sand, sandy soils on river-beds.

In India, muskemelon is grown in Bihar, Uttar Pradesh, Punjab and Rajasthan, while in the south in Andhra Pradesh, Karnataka and Tamil Nadu. In these states, river bed cultivation forms more than 80 per cent of the total area planted under melons in the country, but there are also well-defined pockets in different states where muskmelons are cultivated, such as land and garden areas, in crop rotations and as mixed crop.

Varieties

The popular varieties grown in various parts of the country are Lucknow Safeda, Kanpuria Jogia, Mathuria, Kajra, Jaunpuri in Uttar Pradesh, Sankeda, Galteshwar in Gujarat, Kharri, Jalgaon in Madhya Pradesh, Sanganer, Hara Gola in Rajasthan, Barthesa, Papaya in Andhra Pradesh, Kadapa in Karnataka, Jam in Maharashtra and Hari Dhari in Punjab. The improved varieties are Hara Madhu, Arka Jeet, Arka Rajhans, Annamalai, Durgapur Madhu, NDM 1, NDM 2 etc.

Harvesting and Marketing

Variety climatic conditions, marketing facilities etc., determine the harvesting of muskmelon.

Problems

(1) The best strains in muskmelons are gradually being lost because the cultivators have not much knowledge of seed production.

(2) Attention paid towards breeding varieties is resistant to diseases and insect-pest are minimum.

As there is a considerable scope of increasing the productivity of muskmelon through the production of superior hybrids, there is the need to have it. The future research and development activities should aim at filling the above said gaps and also in developing high-yielding hybrids. Along with these, there is also a need to develop the suitable varieties for particular situations like river-bed cultivation and for late sowing.

WATERMELON

Watermelon (*Citrullus vulgaris*) is a native of tropical Africa, where it has long been used by the wild tribes. The great explorer David Livingstone has mentioned that he saw the countryside near the Kalahari literally covered by water melon vines. It has long been cultivated by the people on the coast of Mediterranean sea. It came to India by the fourth century A.D., Susruta, the great Indian physician who wrote *Susruta Samhita* mentions that this fruit was grown along the banks of Indus river. It was taken to China in AD 900-1000.

Values and Uses

The fruit has hardly any nutritive value as it is made up of nearly 95 per cent water. The balance of 5 per cent consists of fibre 3.3 per cent, a little amount of proteins, far and minerals like calcium, iron and phosphorous with traces of vitamins. It has hardly any calorific value.

Dessert, beer, fresh juice or the juice boiled down to a heavy sweet syrup are various forms in whch the fruit is consumed. The rind is pickled or turned into a sugar candy. Some housewives prepare a spicy durable preparation after grating the rind and drying it. It is fried before serving. In Hawai, it is often used to make pickle. But mostly, it is discarded and used as food by stray cattle flocking near watermelon vendors. The seeds are protein rich with a high edible oil content. The dried roasted seeds are chewed by Chinese in generous quantities. Arabs love watermelon too much.

Dehulled seeds are used for decorating sweets and added to soft drink made out of almond seeds and sonf. They are sometimes grouned and baked in bread.

Watermelon is now grown largely in India, China, Turkey in Asia, Egypt, Algeria and Tunisia in Africa, U.S.A. and Mexico in America and Bulgaria in Europe. Turkey is the largest producer of watermelon in the world. The total production of watermelon in 1997 was 46135 thousand metric tonnes of which Asian region alone contributes 35730 thousand metric tonnes as can be seen from Table 6.6.

Table 6.6 : Production of Watermelons in different regions of the World
(in '000 M.T.)

Region	1995	1996	1997	1998	1999
Africa	2681	2691	3405	3160	3242
N.C. America	2489	2736	2671	2304	2304
S. America	1502	1509	1465	1482	1482
Asia	30506	34337	37415	38140	38151
Europe	3911	3655	3903	3772	3793
World	41172	45019	48956	48968	49082

Source : FAO.

Watermelon in India

Traditionally the cultivation of watermelons was confined to river beds of the Yamuna, Ganges, Narmada in the north and Pennar, Kaveri, Krishna, Godavari in the South but now they are grown in almost all areas. It is a fast growing cash crop for poor and marginal farmers with little area of their own. The total area under watermelons in India in 1998 was 17300 hectares and the production was 2,25,000 metric tonnes. Since 1990 both in terms of area and production, it has been increasing in India as shown in Table 6.7. Now, the cultivation of this has spread to Western and northern Karnataka and Southern Bengal especially the Diamond Harbour.

Table 6.7 : Area and Production of Watermelons in India (1990-1999)

Year	Area (in ha)	Production (in M.T.)
1990	16000	200000
1991	16194	205900
1992	16300	208000
1993	16500	210000
1994	16800	215000
1995	17000	220000
1996	17300	225000
1997	19000	250000
1998	19000	250000
1999	19000	250000

Source : FAO.

Varieties

A very large numbr of varieties ar grown in India. They are Noorjahan, Anarkali, Asahi Yamato, Sugar Baby, Arka Jyoti, Honey Cream, Pusa Rasaal etc. Arka Manik carries high resistance to anthracnose, powdery mildew and blossom and rot and has become popular in Karnataka, Andhra pradesh and Maharashtra. In South Bengal and Maharashtra 'Sugar Baby', an American variety is very popular.

Marketing

The assured markets for this are big cities. In its marketing, the commission agents and travelling traders play an important role. As watermelon is perishable, most of the growers market it at lower price either to the commission agent traders or to the pre-harvest contractors. A very little volume is exported from India to the Arabian countries.

Problems

(1) Lack of organized methods of seed production and distribution of these to the growers.

(2) Problem of transportation and marketing.

(3)   Lack of systematic studies on cost-benefit aspects.

(4)   Problem of insect-pests and diseases.

As watermelons have a great export market especially in the Middle East and internallly the demand is increasing, efforts are needed to overcome the above said problems. There is the need to increase the productivity and production in the country so as to cater the needs of the market. Research and developmental efforts are required to develop breeding programmes for evolving watermelon varieties resistant to spotted wilt virus, Fusarium wilt, anthracnose, watermelon mosaic virus and nematodes. There is also the need to promote ecogeographical surveys of wild relatives and races of watermelon.

Future of Cucurbits

The future of these crops in India is bright since they ar nutritous and also popular vegetables. Both the domestic and external demand for theser are ever increasing. So as to have a better future in this sector, there is the need to supply seeds of pure varieties to the farmers, provide the required pre and post harvesting education and training to the farmers, better marketing, storage and other facilities, timely supply of basic inputs etc., are basically needed. In this regard, a proper preplan is an essential requirement in our country, then only the country can export these in large quantum and may fetch the much needed foreign exchange.

7

Leafy Vegetables

The leaves of a large number of wild and cultivated plants are used as vegetables all over India especially in th rural areas. These vegetables are very rich in minerals and vitamins A and C and they supply the roughage required in our daily diet. These are grown throughout the year, of which some are suitable for winter season and others are suitable for summer. The leafy vegetables comprise some 700 species of plants belonging to 125 plant families. Some of the familiar varieties of these green vegetables are discussed in brief in this part.

AGATHI

It is a tropical tree values for its leaves and flowers. This is a quick growing tree and has both food and fodder values. Its leaves are rich in protein, Vitamin A, Vitamin C and Vitamin B complex. The flowers of it contain protein. Its regular consumption is recommended for goitre patients.

AMARANTH

It belongs to the family Amaranthaceae and the genus Amaranthus. It is grown throughout the year in India since the production cost is low and the productivity is high. The leaves and tender stems are rich in protein, minerals and vitamins A and C. It has great potential for combating cunder and mal-nutrition. It is also called as the 'poor man's spinach'.

INDIAN SPINACH

It is grown throughout India for its tender leaves. The tender shoots and leaves, leaf stalks and stem are chopped and cooked as vegetables.

CURRY-LEAF TREE

It is an under-exploited perennial leafy vegetable, common in backyards of South Indian homesteads. Its aromatic leaves are used to flavour and season foodstuffs. Leaves are also consumed raw in the form of *chutney*. They are added to almost all the curries, hence the name 'curry leaf'. The leaves of it are a good source of carbohydrate, protein and fat and are reputed as a very rich source of Vitamin A and minerals.

DRUMSTICK

Drumstick tree or Moringa is one of the most valuable perennial vegetables. The vitamin rich, mineral-packed nutritious drumsticks chiefly valued for the tender pods, which are cut into pieces and used in culinary preparations. Its leaves are used as vegetables. The fruits of it are good sources of Vitamin B and C; the leaves are rich in Vitamin A and C.

BEET-LEAF

This is the most common leafy vegetable grown in India. It is grown for its tender succulent leaves, however, the tender seed stalk is also cooked. It can be grown in any type of soil having sufficient fertility and proper drainage system. It is of high nutritive value.

CELERY

The native habitat of celery extends from Sweden to Egypt, Algeria and Ethiopia, and in Asia to India, Caucasus and Baluchistan. In India, this is not commercially important as a vegetable crop. However, it is cultivated both for *salad* and seed-raising in the north-west Himalayas, the Punjab, Uttar Pradesh in India. France and U.S.A. are the other major growers of this in the world.

It is a moisture loving plant, requiring a cool climate. In the colder climates and on the hills, it is biennial crop while in the plains it becomes an annual. Celery as a *salad* crop is mostly grown in kitchen.

Uses

Its leaf stalks or petioles are eaten as *salads*, in soups, in sauces, in puree, fried and spiced. The dried ripe fruits are used as spice. The leaves are used as a pre-dinner appetizer as leaves are more nutritious than stalk particularly from the view point of protein, vitamins A and C. The seeds are stimulant and the tonic, is used in Asthma and liver diseases. The seeds are the main ingredients of celery tonic. It has also industrial value. The oil of the seeds is used in the preparation of perfumes and in the flavouring of different kinds of foods as it imparts a warm, aromatic and pleasing flavour to food products.

The total production of celery seed in India is above 5000 tonnes. India exports celery seeds to U.S.A., Canada, France, Japan, Australia, U.K., etc. U.S.A. is the main market for Indian celery seed. The export of celery seed during 1998-99 has increased to 3,991 tonnes value at Rs. 9.69 crores from 3,317 tonnes valued at Rs. 7.99 crores in 1997-98.

CORIANDER

It is a native of the Mediterranean region. Coriander is a cool season crop and is susceptible to milk frost. Coriander seed and fresh leaves are well known spices. The fresh stem, leaves and fruits of coriander have a pleasant aromatic odour. The entire plant, when young is used in preparing *chutneys* and suaces and the leaves are used for flavouring curries and soups. The seeds are extensively exployed as condiment in the preparation of curry powder, pickling spices, sausages and seasonings. It is actually the housewife's secret of tasty dishes. It is used for flavouring pastry, buns tobacco products etc., it is used in the preparation of medicines.

Today, coriander is commercially grown in India, Moracco, Russia, Hangary, Poland, Rumania, U.S.A., Mexico, Guatemala, Bangladesh and Pakistan. In India, it is mainly grown in Rajasthan, Madhya Pradesh, Andhra Pradesh, Tamil Nadu and Karnataka. The total area under this is about 522 thousand hectares and the production is around 308 thousand tonnes.

India exports coriander seeds and coriander powder mainly to Malaysia, Singapore, U.A.E., U.K., Sri Lanka and U.S.A., in large volumes. The export of it was 20,685 tonnes valued at Rs. 46 crores in 1998-99 which was lower than that of 23,734 tonnes valued at Rs. 64.35 crores in 1997-98.

FENUGREEK

It is a native of South Eastern Europe and West Asia and is grown in india too. The leaves and young pods are used as vegetables and the seed as condiment.

The fresh tender pods, leaves and shoots which are rich in iron, calcium, protein, vitamins A and C are eaten as curried, vegetable since ancient times in India, Egypt etc. As a spice, it adds to the nutritive value and flavour of foods. In Egypt and Ethiopia, it is a popular ingredient of bread. In Greece, the seeds, boiled or raw are eaten with honey. In the middle ages, it was recommended as a cure for baldness in men. The powder made from the seeds is used in the Far-East as yellowish dye. The oil is used in the perfume industries.

Apart form India, the other growers of fenugreek in the world are Argentina, Egypt and the Mediterranean countries. The total area under fenugreek in India is around 50,000 hectares and the production is about 60,000 tonnes. Rajasthan stands first both in terms of area and production followed by Gujarat, Haryana and Uttar Pradesh.

India exports fenugreek seed and powder to Y.A.R., U.A.E., U.K., Saudi Arabia, Sri Lanka, South Africa, Japan and Netherlands. However, the volume of export is just 10 per cent of the total internal production. In 1998-99 the export of fenugreek increased from 6,006 tonnes valued at Rs. 9.87 crores in 1997-98 to 10,082 tonnes valued at Rs. 19.15 crores in 1998-99.

The above are some of the important leafy vegetables as well as condiments grown in our country. There is no database of reliable statistics on all of these leafy or green vegetables as far as area, production etc. are concerned. As a leafy vegetable, most of them are grown in the kitchen gardens. However, there is vast scope for commercial exploitation of these, which demands real attention. There are some major problems in these, like the problem of transportation, storage, marketing etc. Apart from these, the available varieties are traditional, hence efforts are needed to improve the pre- and post-harvest technologies with regard to these leafy vegetables in our country.

8

Legume Vegetables

As sources of food, Legume vegetables are next in importance to cereals. They contain more protein material than any other vegetable product, carbohydrates and fats are also present in these. The legumes or pulses all belong to the great family of Leguminosae, which is characterised by having a special kind of fruit, a legume, which is a pod that opens along two sutures when the seeds are ripe. Nearly 11,000 species of legumes are known, and many are of importance as industrial, medicinal or food plants. They have been cultivated and used for food for centuries all over the world. Legumes are easily grown, mature rapidly, and are highly nutritious. They are rich in proteins, minerals and vitamin B. Before the advent of potato, they constituted a great part of the food of the poorer classes in Europe. Legumes have a high energy content and are particularly well adapted for use in cold weather or where physical extension is involved. The immatured fruits are served as a food.

There are at least 18 types of cultivate beans. From the standpoint of green vegetables French bean, Green peas, Cowpea, Cluster bean, Field bean are the most important. The different types of beans vary greatly in flavour, season and other characteristics.

PEAS

Pea (*Pisum sativum*) is a native of Southern Europe and has been cultivated before the beginning of the Christian era. Peas were well known to the Greeks and Romans. Peas are annual, glaucous, tendril-bearing, climbing or trailing plants, with white or coloured flowers and pendulous pods. There are two major groups of peas. Field peas and Garden peas. Field beans are grown for the seed, which are used for human consumption in the form of pea meal or split peas. The plants are used for forage and green manuring while Garden peas contian more sugar than field peas, and the seeds are eaten green or are used for canning purposes.

India is the larges producer of green peas in the world followed by U.S.A., China, France and U.K. The world production of green peas in 1999 was 6892 thousand M.T. which was stagnant over the years as can be observed from Table 8.1. The total area under this in the world in 1999 was 821 thousand hectares.

Table 8.1 : Production of Peas in Major Producing Countries in the World
(lack M.T.)

Country	1995	1996	1997	1998	1999
India	21.50	21.50	20.00	20.00	20.00
U.S.A.	11.12	9.39	10.89	10.97	10.97
China	7.39	9.19	10.16	10.16	10.16
France	5.57	5.75	5.15	5.54	5.50
U.K.	5.35	5.35	4.49	4.17	4.17
World Total (Including Others)	71.83	70.97	68.76	68.98	68.92

Source : FAO.

Area and Production of Green Peas in India

In India, peas are grown mainly in Uttar Pradesh, Bihar, Haryana and Punjab. Uttar Pradesh alone accounts for 70 per cent of the total output of peas in India as can be observed from Table 8.2. As a whole, the area under green peas in India is about 200 thousand hectares and the production is more or less stagnant at around 200 thousand M.T. as can be noted from Table 8.3.

Varieties

Green pea varieties are grouped into two, smooth-seeded and wrinkle-seeded. Several varieties are recommended by the IARI in India which are given below :

(1) Early Varieties

These are Arkel, Jawahar Matar 3, Jawahar Matar 4, Harbhajan, PM2, Jawahar Peas 54 etc. These varieties are now-a-days getting more popular because of better economic returns. It is not that they yield more but the initial price fetched makes them

Table 8.2 : State-wise Area, Production and Yield of Peas

(Area : '000 ha, Output : '000 M.T., Yield M.T./ha.)

State	1993-94			1994-95			1995-96		
	Area	Production	Yield	Area	Production	Yield	Area	Production	Yield
Assam	30,694	16,097	0.52	4,240	1,376	0.32	4,245	1,275	0.30
Bihar	4,000	32,000	8.00	14,800	1,69,200	11.43	15,400	1,84,800	12.00
Haryana	6,450	83,000	12.87	6,900	99,000	13.19	7,000	93,450	13.16
Himachal Pradesh	7,107	70,228	9.88	7,220	71,800	9.94	7,270	72,773	10.01
Karnataka	1,295	16,187	12.50	2,259	28,238	12.50	2,277	28,463	12.50
Madhya Pradesh	24,500	2,45,000	10.00	1,904	19,000	9.98	1,999	20,000	10.01
Manipur	740	3,700	5.00	792	3,960	5.00	797	3,988	5.00
Orissa	6,300	51,500	8.16	6,350	52,000	8.17	6,540	54,600	8.19
Punjab	13,100	79,070	6.04	13,100	79,070	6.04	13,200	79,701	6.04
Rajasthan	3,630	9,823	2.71	916	1,807	1.97	5,730	13,788	2.40
U.P. (Hills)	10,369	44,749	4.32	10,816	53,301	4.32	10,991	56,066	5.10
U.P. (Plains)	91,228	8,78,530	9.53	66,019	8,17,996	9.63	1,40,468	16,71,872	12.39
West Bengal	3,010	31,094	10.33	3,000	31,000	10.33	3,000	31,000	10.33
Delhi	1,137	8,394	10.80	2,088	12,626	7.38	768	4,771	6.05
Total	1,81,611	15,28,371	8.42	2,18226	23,06,298	10.57	2,23,965	23,41,313	10.45

Source : National Horticultural Board.

Table 8.3 : Area and Production of Green Peas in India
(Area in '000 ha, production '000 M.T.)

Year	Area	Production
1990	145	1900
1991	147	2106
1992	147	2100
1993	148	2150
1994	148	2150
1995	145	2150
1996	148	2150
1997	200	2000
1998	200	2000
1999	200	2000

Source : FAO.

highly suitable for commercial cultivation. There are two pickings of green pods in early varieties. The first picking may be taken in two months of sowing and the second, after an interval of 15 days.

(2) Mid-season Varieties

They are Bonneville, Jawahar Matar 1, Jawahar Matar 2, IP3, P 88, Jawahar Pea 83 etc. These are high-yielding varieties capable of giving three green pickings. The first pick can be taken after 90 days and the subsequent two at an interval of 15 days.

Early planting of peas fetches higher income and the farmers in some districts of Madhya Pradesh plant peas on hillocks during Mid-August, whenever there is break in rains.

Harvesting

The green pods are to be picked before the webbing starts. The picking may start as soon as the green ovules are fully developed and pods still overmature.

Exports

India exports fresh or chilled peas, shelled or unshelled and frozen peas. As far as fresh or chilled and shelled or unshelled peas are concerned, Mynamar, Sri Lanka, UAE are the major markets. During 1997-98 a total volume of 134.87 M.T. was exported and its value was Rs. 19.55 lakhs. While from the exports of frozen peas, India earned Rs. 21.50 lakhs during 1997-98 when it exported 81.59 M.T. The major consumers of these are UAE and USA. As far as non-frozen peas exports, are concerned; Kuwait, Russia, Saudi Arabia and UAE are the major markets. India earns substantial amount from these.

As the popularity of canned, frozen and dehydrated peas is increasing in the world, the scope for an improvement in its production is more in India. In this regard, the prevailing problems like marketing, storage, diseases and insects have to be solved. Efforts are needed to supply high-yielding strains so as to produce better quality pods along. Then only the country can cater the needs of this evergrowing demand both that of domestic as well as external.

FRENCH BEAN

The french bean or kidney beans (*Phascolus vulgaris*) are natives of the new world. They were probably domesticated by the Incas and were early used by the Indians of both South and North America. Now the young pods, the unripe seeds and the dried ripe seeds are used for human consumption, while the whole plant is used for forage. These beans are low, errect annuals with small white or coloured flowers and slender pods. They are grown as either bush or pole beans, and over a thousand varieties are cultivated. The important pole bean varieties grown are Premier, Bayo Brown Swedish, Kentuck wonder etc., while the bush bean includes Pusa Prabati, Giant stringless, Arka Komal Sel-2, Sel-3, Sel-9 etc.

FIELD BEAN

The lablab bean is one of the most ancient among the cultivate plants. In the west, it is known as 'Bonavist' or the 'Hyacinth bean'. It is grown throughout the tropical regions of Asia, Africa and America. It can be used as pulse, vegetable and for forage. It is richer than the french bean in its nutritive value.

In India, the crop is mainly grown for its green pods, while the dry seeds are used in various vegetable preparations. It is a field crop of Tamil Nadu, Andhra Pradesh, Karnataka, Madhya Pradesh and Maharashtra.

The most important varieties grown in India are CO1, CO2, Hebbal Avare3, Wal Konkan 1, 125-36, etc.

As beans are valuable source of protein, calcium, iron and vitamins and are a cheap source of nourishing food there is the need to improve the production and productivity of these. Along with these, an extension in area is also required in our country. For all these, a planned strategy is required for its future. This strategy should include the following aspects. They are :

(1) Comparing the performance of many cultivars under widely differing climates.

(2) Comprehensive germplasm collection.

(3) Identification of suitable Rhizobium strain for inoculation.

(4) Uniformly maturing photo-insensitive varieties resistant to pest, diseases, adverse soils and weather.

(5) Compatibility as an intercrop with food and forage legumes.

(6) Improvement of feeding value of seeds for human and livestock.

(7) Reduction of anti-nutritional factors.

(8) Improving the post-harvest technologies

(9) Exploring the possibilities to increase the exports.

(10) Collection of data on area, production and the related matters.

9
Marketing of Vegetables

The marketing of vegetables is a complex process. It consists of all those functions and processes involved in the movement of the product from the place of production to the place of consumption. The marketing activities involve not only the functions of buying and selling, but also the preparation of produce for marketing, assembling, packaging, transpotation, grading, storage, processing, retailing etc. The number of functions and its type vary from product to product, from time to time and from place to place. The major goals in marketing are :

(1) To meet the domestic requirement of protective foods for the rising population;

(2) To provide raw material base for industry;

(3) To develop appropriate system for reducing post-harvest losses;

(4) To enhace exports.

So as to obtain these, an efficeint marketing system should have the follwoing objectives. They are :

(a) To enable the primary producers to reap the best possible benefits through price mechanism;

(b) To provide facilities for lifting of all produce, the farmers are willing to sell at an incentive price;

(c) To reduce the price spread between the primary producer and ultimate consumer;

(d) To make available all products of farm origin to the ultimate consumer's at reasonable prices without impairing the quality of the produce.

The prevailing marketing system for vegetables in India is constituted of a three-tier network of markets viz., Village markets, Secondary markets and Terminal markets.

Several phases of agricultural marketing practices are undertaken which include :

(1) At the farm gate–after harvesting for covertion into the desired form acceptath!e at the rural or primary market.

(2) At the rural markets-for change of hands for the producers to the village merchants or the agents of the buyers or the local consumers.

(3) At the asembling markets–where the village merchants/ producer-sellers sell the produce to the wholesalers or the commission agents for onward dispatch to the clients in the upcountry markets.

(4) At the secondary markets–where the prodce from the assembling markets in the market shed is stored and transacted for distribution to the terminal market for consumption.

(5) At the terminal markets–for vending through wholesalers and retailers for final consumption or for shipment to other countries.

(6) At the retail level to consumption point which may be a processing unit or an individual consumer.

Packaging of Vegetables

Packaging is an esential requirement in the marketing process of vegetables. It is required becasue, (a) it protects the product against breakage, spoilage etc., during its movement; (b) it facilitates the handling of the product during storage and transporatation; (c) it ensures cleanliness of the product; (d) it prolongs the storage quality of vegetables by providing protection from the ill-effects of weather.

The type of the container used in the packaging of vegetabes varies with the type of vegetable as well as with the stage of marketing. However, in general, gunny bags, wooden boxes or straw baskets are used for packing vegetables in our country. But in these, the cost of packaging is more and the scope for spoilage is also higher. In the case of vegetable packaging, the prevailing methods are traditional and the absence of an innovative packaging technology has led to immense losses to the vegetable growers in our country. For a reduction in the wastages, there is the need for a innovative packaging system which can reduce spoilage, considerable savings in overall cost of packaging and lighter in weight leading to lesser damages for transportation etc. In this regard, the efforts of IIM, Ahmedabad, are worth mentioning since it has developed three different Vastapur cartons for tomatoes. These are useful for the growers to send their produce to markets where they can get better price and the cost of packaging is also less on account of reduced spoilage. In several towns, super markets have started supplying vegetables duly cleaned and packed in corrugated fibre board containers both for domestic and export markets. Recently, processing industries and cooperative societies started using plastic crates for transportation of vegetables from the filed as well as for storage, thereby reducing spillage to the maximum extent.

Transportation

The movement of vegetables between places is one of the most important marketing functions at every stage. The vegetables grown have to be brought from the farm to the local markets and from there to the primary wholesale markets, secondary wholesale markets, retail markets and ultimately to the consumers. Transport is an indispensable marketing function since it has advantages like : (a) it widens the market, (b) narrowing price difference over space, (c) creates job opportunities and (d) facilitates specialized farming etc.

In the case of most of the vegetables, the cost incurred by the cultivators towards transportation cost ranges in between 40-50 per cent of the total cost incurred by them due to bulky nature and perishability of the produce. The available means of transport are mostly insufficient and most of the vegetable growers in India dare carrying their produce either by head load or through bullock-carts and to some extent by modern means of transport like jeep or Autos. The non-availability of all weather roads, sufficient road network, quickest means of transport etc., are some of the basic and major problem faced by the vegetable growers in our country which ultimately lead to huge amount of losses or a fluctuation in the supply and thereby the price also.

Grading

Grading and standardization is a marketing function which facilitates the movement of produce. Grading standards for commodities are laid down first and then they are sorted out according to the accepted standards. Garding has certain advantages like it fetches higher prices, creates an awareness, widens the market, reduces the cost of marketing etc.

Agricultural Produce (Grading and Marketing) Act, 1937

This is the first legislation enacted by the Central Government. The Act empowers the Central Government to formulate standards and carry out grading and marking of agricultural and allied commodities. The articles included in the schedule includes vegetables, only the onions, table and seed potatoes.

In the case of vegetables, at present, no grading is done at either the field level or at the market place before the produce is sold, though it was proved by many studies that grading fetches better price for the produce. The main reasons for non-grading of vegetables are : (a) the fear by the cultivators that if the produce is graded and sold, it may become difficult to dispose off the low grade produce, as in the case of co-operative society where low grade produce is not accepted; (b) grading is voluntary; (c) lack of proper standards for standards for different vegetables; and (d) lack of grading facilities at the markets.

COLD STORAGE OF VEGETABLES

Vegetables are highly perishable and have short life. The vegetables are having living tissues and they undergo normal life process. This process leads to a gradual deterioration. If they are not disposed or marketed without any delay. The failure in its disposal leads to a lot of wastage of these products. In India, the estimated loss in th post-harvest activities is ranging in between

20-40 per cent of the total production of vegetables in a year. So, a good deal of nutritious food in the form of vegetables is lost even before it reaches the consumers. So as to overcome this problem or at least to minimise the volume of wastages, it calls for better storage facilities. In India, vegetables are stored in two ways, viz., Home storage and cold storage. In the first method, the vegetables can be stored ony for a very shorter period while under cold storage it may be stored for longer time since it is possible to maintain the desired point of temperature and humidity. With the help of cold storage facilities, it is possible to offer vegetables for consumption or for processing in fresh conditions over a long period.

Importance of Cold Storage

Cold storage is a vital link between the production and consumption of vegetables. Apart from the conservation of vegetables, it also helps in increasing the marketing period and ensures availability to the consumers over a long period. Cold storages also ensure reasonable price to the producer, who does not have to resort to distress sale. He can sell his produce after a reasonable period of storage when the rice is remunerative. The recommended storage temperature and relative humidity for vegetables can improve the life of these as can be observed from Table 9.1.

Cold Storage Industry in India

The Royal Commission on Agriculture had observed as early as in 1928 that the cold storage industry was bound to play a very important role in India's agricultural economy. Although the first cold storage was established in Calcutta in 1892, the industry made tardy progress up to 1950's; the progress up to 1955 was very slow and there were only 83 cold storages with an installed capacity of 42,965 tonnes. During that period, it was felt that a lot of quantitative and qualitative improvement was necesarry in cold storages. It was also observed that the cold storages were not established on technically sound lines and the storage conditions prevailing there were not proper and hygienic. It was, therefore, felt necessay to regulate the industry to provide for the technical specifications and minimum norms for hygienic storage. The need

Table 9.1 : Recommended Storage Temperatures and RH for Vegetables

Vegetables	Temperature (ºC)	RH (%)	Storage Life (Weeks)
Asparagus	0.0	95	3-4
Brinjal	10.0-11	92	2-3
Doliches lablab, pod	0.6-1.7	90	3
Beet, topped	0.0-1.7	90-95	2-14
Beet, Bunches	0.0	90	1.5
Bitter guard	0.6-1.7	25-90	4
Cabbage, early	0.0-1.7	92-95	4-6
Cabbage, late	0.00-1.7	92-95	12
Carrot, topped	0.0	95	20-24
Cauliflower (Snowball)	0.0-1.7	25-95	7
Celery	0.6-0.0	92-95	8
Colocasia	11.1-12.2	25-90	21
Coriander, leaves	0.0-1.7	90	5
Cucumber	10.0-11.7	92	2
Garlic	7.2-10.0	75	16-24
Ginger	7.2-10.0	75	16-24
Lettuce, head	0.0	90-95	3
Lettuce, leaves	0.0	95	1
Lime bean, pods	4.4-7.2	90-95	15-2.0
Muskmelon	1.7-3.3	85-90	1.5
(Honey dew)	7.2	25	4.5
Okra	0.9	90	2.0
Onion, leaves	0.0	90-95	2.0
Onion bulbs, red & white	0.0	70-75	20-24
Pea, green	0.0	22-92	2-3
Pepper, green	7.2	25-90	3-5
Pepper, ripe	5.6-7.2	90-95	2
Potato (Irish)	3.0-4.4	25	34
Pumpkin	1.7-11.6	70-75	24-36
Reddish	0.0	22-92	3-5
Squash, winter	12.2-15.6	70-75	24-36
Sweet corn	0.6-1.7	90-95	1
Sweet potato	10.0-12.2	20-90	13-20
Tapioca, root	0.0-1.7	25	23
Tomato, unripe	8.910.0	25-90	4.5
Tomato, ripe	7.2	90	1
Turnip	0.0	90-95	2-16
Watermelon	7.2-15.6	20-90	2

for introducing a licensing system for the cold storage was also stressed by the Horticultural Development Board as early as in 1957. The Ministry of Agriculture, Government of India, thereafter promulgated the Cold Storage Order, 1964, under section 3 of the Essential Commodities Act, 1955, which came into force with effect from January 1, 1965. It is being administered by the DMI to achieve the following objectives :

(1) to ensure hygienic and proper refrigeration conditions in the said storage.

(2) to regulate the growth of the cold storage industry in a planned manner.

(3) to render technical guidance for scientific preservation of food stuffs; and

(4) to safeguard the interests of farmers and depositors.

According to the provisions of the order, it is obligatory on thepart of every cold storage owner to obtain a licence from the licensing officer for storage of food stuffs like fruits, vegetables, meat, fish, dairy and poultry products etc. This order is applicatble all over the country except Uttar Pradesh and West Bengal, where State Governments have enacted their own cold storage legislations. In 1979, Punjab and Haryana were also allowed to promulgate their own orders. At present, the DMI has implemented the Cold Storage Order, 1964 (Revised 1980) all over the country except the state of Punjab, Haryana, Uttar Pradesh, Bihar and West Bengal to ensure hygienic and proper refrigeration conditions for the storage of perishables and regulate planned growth of cold storage capacity. It was repealed in 1997 with a view to remove licensing and attract private investment as a move of economic liberalisation. However, the five states having their own legislation are continuing their implementation. In the 1999-2000 budget, a credit-linked capital subsidy scheme for the construction of cold storage and godowns to be implemented by the Ministry of Agriculture with NABARD was proposed. A plan is on the table to create an additional cold storage capacity of 12 lakh tonnes and modernisation/upgradation of the existing 8 lakh tonnes capacity. An additional capacity of 4.5 lakh tonnes for the storage of onions alone is also proposed to be created.

Apart from the above, the Central Warehousing Corporation has plan to set up a controlled atmosphere storage at Sirhind with a capacity to handle 8000 tonnes of fruits and vegetables, under a storage utilisation tie-up with the Punjab Agro Indsutries for the storage of exotic fruits and vegetables. Above all, the APEDA has set up cold storage at the Indira Gandhi International Airport in New Delhi and Bangalore airport. It is also selling up 3 more cold storage facilities, one each near international airport in Chennai, Thiruvananthapuram and Hyderabad. APEDA also has plans to set up cold storage facilities at Amritsar, Ahmedabad and Kolkata and walk in type of cold storages at Guwahati and Agartala to promote the exports. The Maharashtra Government is setting up two air cargo terminals one each in Pune and Nagpur for handling the export of perishables.

In general, the real interest on cold storage started after the launching of "Grow More Food" campaign in the First Five Year Plan when it attracted the attention of the industrialists, growers and consumers, and the number of cold storages rose upto 359 with an installed capacity of 0.31 million tonnes by the end of the year 1960. During the third five year plan, the government provided incentives for setting up of more cold storages, especially for storing potatoes. From then onwards, the number of cold storages increased slowly and steadily as can be observed from Table 9.2.

Table 9.2 : Cold Storage Industry in India

Year	Number	Storage Capacity (in Million tonnes)
1947	4	0.003
1955	83	0.04
1960	359	0.31
1965	600	0.68
1970	1,218	1.64
1975	1,615	1.99
1980	2,283	3.96
1988	2,659	5.58
1990	2,930	7.68
1995	3,253	8.73
1999	3,443	10.30

Source : DMI, Faridabad.

The total number of cold storages at present is 3,443 with an installed capacity of around 10.3 million tonnes; out of these 3,443 cold store units, 2975 are in the private sector and 303 in the co-operative sector and the rest in the public sector.

Defects

(1) The relative utilization of the existing capacity for commodities shows that nearly 90 per cent of the cold storage capacity is being utilised for the storage of potatoes alone while for fruits and vegetables the share in the total is just one per cent.

(2) Most of the cold storages are under the control of the privateers. As the NCA observed that preponderance of private sector in this industry has given rise to a number of malpractices like overcharging, hoarding of storage space etc.

(3) As far as sector-wise distribution of ownership of cold storage is concerned, nearly 86 per cent of the total numbers are under private ownership, while the cooperative sector accounts for about 9 per cent and the remaining under the public sector. This excess involvement of the private sector leads to the expoitation of the producers by charging an exorbitant rents for storage during the peak season.

(4) The available capacity of cold storages is insufficient as far as the total production of vegetables is concerned. It is observed that the capacity of cold storage and frozen storage in India is barely suffecient to meet 14 per cent of potato produciton, 0.5 per cent of fruits and vegetables and 0.47 per cent of fish. The country is deficient in cold storage facilities upto 20 per cent of potato output that requires to be stored for sale in the lean period.

(5) Most of the cold storage facilities are concentrating in the terminal markets and they neglect the production centres Due to this negligence, a huge amount of loss is taking place.

(6) Lack of improvements in the traditional methods of storages.

So, as to avoid the prevailing loss in the vegetable sector and to enhance the production and marketable surplus, there is an urgent need to formulate a long term strategy for cold storages in the country. This strategy should concentrate on the following aspects :

(1) Installation of cold storage facilities according to the production of vegetables should be undertaken at each production centre. Side by side, efforts are also needed to promote the existing traditional storage methods in rural areas.

(2) The state should come forward to have separate or independednt State Cold Storage Acts so as to increase the efficiency of cold storages and to minimise the wastes. Efforts are also needed from the side of the state governments to supply the cold storage facilities to each cooperative societies which can minimise loss of vegetables on the one side and on the other it may lead to an increase in the production of vegetables.

(3) There is an urgent need to control the activities of the private sector cold storages so as to safeguard the interest of the growers and to minimise the fluctuations in the suppy as well as the prices.

(4) Efforts are needed to start the mobile cold storage facilities in the growing areas. In this regard, certain incentives like financial assistance, subsidies etc. will be useful.

(5) Adequate capacity should be created under the public and cooperative sector so as to provide competition to the private cold storage owners.

(6) Sufficient space should be created in the international airports for the vegetables under the cold storage facilities so as to expand our exports.

(7) Need to modernise/upgrade the existing cold storage facilities.

In spite of the developments in the cold storage facilities in our country over these years, the prevailing situation is not a perfect one when it is compared with the growth rate of vegetable production. Because of this, the prices of vegetables frequently fluctuate and the living standards of the consumers are disturbed. To be highly effective in controlling the wide seasonal fluctuation in the prices of vegetables and to have a proper supply to the internal and extenal markets, there is the need to increase the capacity of our cold storages. Hence a long term planned strategy on scientific lines is esentially needed which will be useful and helpful in attaining the recommended per capita consumption of vegetables and also in improving the earnings of the growers. So as to improve our export of vegetables, there is an urgent need to improve the quality of cold storages, as there is much talk about the efforts being made for globalisation and liberalisation to open up the economy for attracting more and more foreign investment particularly for infrastructure building. In this direction, we have signed WTO and SPS agreements which entail following international/CODEX standards of quality etc. In this regard, there is the need for a cautious and balanced approach in the cold storages of the State Governments which are covering about 70 per cent of the total cold stores in the country.

PROCESSING

It is an important marketing function in the present day marketing of vegetables. The popularity of value added products has made the way for an improvement in processing technologies and most of the developed countries are concentrating in it so as to cater the needs of the consumers. In India too, the processing of vegetables has gained importance in recent years. Processing activity involves a change in the form of the commodity and is concerned with the addition of value to the product by changing its form. Several advantages are there in this, like it increases the total revenue, reduces the losses or wastages, it is possible to store for a long period, it generates employment opportunities and it widens the market.

Vegetable Processing Industry in India

Most of the vegetables in India are still consumed fresh except for a very small quantity going into the manufacture of pickles, tomato ketchup, dried and fried potato, cassava and sweet potato chips. The production of frozen peas, garlic and ginger paste, tomato puree etc. has been taken up only in recent years in India. Even though India is the second largest producer of vegetabels in the world, only 0.5-1 per cent of the total vegetables produced are processed as against 83 per cent in Malaysia, 80 per cent in South Africa and 65 per cent in the USA. The installed capacity for processed fruits and vegetables in India wet up from 275 thousand tonnes in 1981 to 950 thousand tonned in 1991 and 1910 thousand tonnes by 1996.

As far as the number of units registered under Food Products Order, 1955, which covers units in the small scale sector also went up from 1996 in 1981 to 4,700 by 1996. In terms of production, it was 90 thousand tonnes in 1981 and went up to 974 thousand tonnes in 1996. At present, the capacity utilisation by these units is around 50 per cent. This is mainly due to the seasonal nature of the industry on the one hand and the lack of diversification and aggressive marketing on the other.

As far as the scales of these processing units are concerned, the maximum number of processing units are in the home scale (43 per cent), minimum being in large scale (9 per cent) sector and about 30 per cent are relabellers. It is interesting to note that the large scale units contribute to about 70 per cent of the total production. In terms of ownership, majortiy of these units are in the private sector, which accounts for about 95 per cent, and the balance being distributed among the public and cooperative sector As far as the locations of these units are concerned, 41 per cent of them are located in the Western region, 28 per cent in the Northern, 22 per cent in the Southern and 9 per cent in Eastern region.

The popular procesed foods produced in the country include potato chips, dried onions and garlic powder, ginger and ginger paste, canned beans, frozen peas, cauliflower and okra, tomato ketchup and puree, chilly sauce etc. Of the total production of processed vegetables about 40 per cent is consumed through the hotels, defence services, in railways and airlines and 40 per cent is purchased by the housewives and households, and the remaining 20 per cent are exported in terms of vaue. Th produciton of frozen vegetables has gone up in recent years for the domestic as well as for exports also. Free dried and individually quick frozen vegetables are being produced in India mainly for exports. Several 10 per cent export oriented units are engaged in the production of these and there are some multinationals like Pepsi Foods, VST Natural Products Ltd., etc. in the production of pickled vegetables and sliced cucumber.

The above aspects clearly show that the procesing of vegetables in India has not developed to the desired extent, hence, the scope for improving this is vast, since the demand in the domestic as well as external markets is even increasing in recent years. So, efforts are needed in this regard which will minimise the volume of wastages and can generate more income and employment in our country. For having these, there is the need to develop low cost technologies and to operate them at farm level on a cooperative basis which will help in better utilization of waste, thereby reducing the pollution load in cities where processing factories are functioning. In India, vegetable processing has immense potential and a number of units can be set up in a small, medium and large scale sectors in producing centres based on modern technology and possibly with foreign collaboration so that these products are manufactured on a mass scale and marketed both in the domestic and export markets. So as to develop this sector, there is the need to create the required infrastructural facilities.

Marketing Channels

There are the routes through which vegetabels move from the place of production to the place of consumption or from producers to consumers. The length of this varies from product to product, depending on the volume to be moved, the form of consumer demand and degree of regional specialization in production. The marketing channels for vegetables, in general, vary from commodity to commodity and from producer to producer. The small farmers usually sell their produce either to the village traders or in the village markets while large scale farmers sell their produce in the main market, where it goes into the hands of wholesalers or commission agents. In India, most of the vegetable cultivators are small farmers and they usually sell their produce in the weekly village markets or shandies either to the commission agents or to the consumers. Sometimes in the rural areas and small towns, many producers themselves perfer the functions of retail sellers The common and typical marketing channels for vegetables found in India are :

(1) Grower-Consumer.

(2) Grower-Retailer-Consumer.

(3) Grower's Co-operative-Consumer.

(4) Grower-Grower's Co-operative-Commission Agent- Retailer-Consumer.

(5) Grower-Forwarding Agent-Commission Agent-Retailers- Consumer.

(6) Grower-Grower's representative-Retailer-Consumer.

(7) Grower-Wholesale Merchant-Retailer-Consumer.

(8) Grower-Wholesale Merchant-Commission Agent-Retailer in distributing market-Consumer.

(9) Grower-Pre-harvest contractor-Commission Agent in assembling market-Commission Agent in distributing market-Retailer-Consumer.

(10) Grower-Commission Agent-Wholesaler-Retailer- Consumer.

In the above system, there is a multiplicity of interaction and involvement of a large number of market functionaries/ intermediaries with conflicting interests. The prevailing marketing system is traditionally dominated by the traders. The producer seller continues to be the weakest link in the chain. This system is unfavourable to the farmers still this is there because of the absence of infrastructure and improper management coupled with lack of market intelligence etc. As far as the intervention of the government is concerned in the marketing of vegetables except those of onions and potatoes, it is nil.

EXPORT OF VEGETABLES

In spite of high domestic consumption, there is good export potential for various fresh, canned and dehydrated vegetabels, including chillies, onion, garlic, okra, tomato, potato, asparagus, broccoli, pumpkin, carrot, celery, brinjal, cabbage, cauliflower, beans and peas. Among the fresh vegetables exported, the principal ones are Okra (60%), bittergourd (20%), chilli (10%) and mixed vegetables (10%). The traditional items of our exports include onions and potatoes also. The APEDA has identified some of the non-traditional vegetables such as asparagus, celery, sweet pepper, sweet corn, baby corn, green peas, french beans and tomato cherry etc. which have good export potential. The traditionally exported vegetebles have markets mainy in Malaysia, Singapore, the Gulf countries, Sri Lanka and Bagladesh. The non-traditional vegetables have export marketes mainy in Europe and South East Asia. In South-East Asia, the demand for the vegetabels exists throughout the year while in Europe it is normally between November to May only when because of extreme winter conditions, nothing grows there.

The export of fresh vegetables takes place principally from Mumbai, Delhi, Bagalore and Trivandrum. The Delhi based exports are large from April to August which include okra, *lanki*, *tinda arvi* and *kachla* to Kuwait, Dubai and Saudi Arabia. From September to February radish, carrot, green peas, cauliflower, cabbage, green chilly, ginger are exported. These two principal export months overlap in March. There is also one single peak period during the 40 days of Ramzan when the market can absorb a maximum volume per week. During this period, the scope for doubling exports is there but non-availability of cargo space restricts this.

The vegetables exported from Mumbai are procured from Maharashtra, Gujarat and Kerala. However, the entire volume of okra comes from Maharashtra alone. The markets in Mumbai are located at Dadar, Matunga, Byculla and Mahar. Of the total vegetables arriving to these markets only 0.1 per cent are exportd. The Delhi exporters secure their requirements from the Azadpur Mandi, where the vegetables are stored in hired space.

Export form Delhi unlike those from Mumbai include all varieties almost throughout the year. The important vegetables exported from this centre are French beans, *imli*, *tinda*, *suran*, *parwal*, brinjal, *surti* beans, *toria*, *arvi*, tapioca, okra, bittergourd, chilli etc. For exports, packaging is almost entirely in corrugated boxes and is normally done in 5 kg to 10 kg. unless specified

In 1996-97, the country exported fresh vegetables to the tune of Rs. 341.15 crore over Rs. 301.19 crore in the preceding year, representing a growth of 11.71 per cent. Howevr, it came down to Rs. 319.44 crore in 1997-98 representing a negative growth of 6.36 per cent, which was mainly due to a decline in the earnings from onions while that of other vegetables it has increased sharply as can be observed fom Table 9.3. An analysis of India's exports to principal markets reveal that the UAE, Malaysia, Sri Lanka and Bangladesh together accounted for a predominant share of above 66 per cent of the total exports of fresh vegetables. As far as item-wise vegetables are concerned, next to onion, potato, mixed vegetables, tomato, garlic dominates as can be observed form Table 9.4 where figure are given for the period 1994-95 to 1997-98.

Table 9.3 : Export of Vegetables from India (Rs. Crores)

Year	Onions	Other Vegetables	Total
1994-95	204.61	44.13	248.29
1995-96	230.72	70.47	301.19
1996-97	265.21	75.94	341.15
1997-98	205.37	114.07	319.44

Source : APEDA.

As far as the export of processed vegetables are concerned, the main items exported form India are dried and preserved vegetables, chilli pickles, green pickles, tomato chutney and paste, soya sauce etc. Tomato ketchup and sauce, chilly sauce etc. Table 9.5 shows our performance in the export of some of the major processed products over the years. For canned vegetables, the major markets are the UK, USA, Saudi Arabia, FRG and UAE and for dehydrated vegetables Russia, Japan, U.K. and FRG while for pickles and *chutneys*, the main markets are the U.K., Saudi Arabia, the USA and UAE.

India exported Rs. 472.96 crores worth of dried/preserved vegetables in 1997-98 which was double the amount of Rs. 231.08 crores that of the previous year. Through the export of pickles/ *chutneys*, the country earned Rs. 79.67 crores in 1997-98 which was Rs. 56.43 crores in 1996-97.

Table 9.4 : Exports of Some Fresh/Chilled/Dried Vegetables from India

(Quantity : M.T. Value Rs. Lakhs)

Description	1994-95		1995-96		1996-97		1997-98	
	Quantity	Value	Quantity	Value	Quantity	Value	Quantity	Value
Potatoes	15,755	669	34,516	1,890	24,936	1,716	20,884	904
Tomatoes	1,072	63	646	36	690	42	863	41
Onions	4,01,282	20,462	3,50,989	20,720	4,27,012	26,521	3,33,549	20,264
Shallots	714	44	166	13	992	56	200	10
Garlic	422	43	3,524	333	3,650	387	2,437	220
Leeks	1	-	-	-	-	-	48	2
Cauliflower and Broccoli	25	1	-	-	12	2	49	7
Kohrbi	135	4	34	1	54	2	26	6
Cabbage lettuce	18	-	-	-	209	3	12	-
Carrots and turnips	-	-	33	3	1	-	70	8
Other roots	14	1	143	5	16	2	82	5
Cucumber and gherkins	1,101	178	5,015	646	9,609	1,449	10,766	1,784
Peas, shelled or unshelled	17	68	556	123	136	35	135	20
Beans, shelled or unshelled	-	-	69	15	1	-	121	27
Globe artichokes	-	-	-	-	-	-	140	3
Mushrooms	290	174	1,560	640	2,344	861	5,711	2,013
Green chilli	46	840	142	22	1,093	167	869	120
Other chilli	117	17	51	11	223	75	246	35
Olives	7	-	2	3	-	-	15	3
Plantain (curry banana)	20	3	152	33	20	1	38	3
Mixed vegetables	8,591	749	14,280	1,210	10,318	917	17,957	1,923
Green pepper	19	5	73	12	203	49	375	167
Pumpkins	35	1	48	1	88	3	147	4
Other vegetables	6,127	476	7,182	557	6,182	568	30,180	3,226

Source : DGCIS.

Table 9.4 : India's Exports of Pickles, Chutneys, Sauces and Pastes

(Quantity : M.T. Value Rs. Lakhs)

Description	1994-95		1995-96		1996-97		1997-98	
	Quantity	Value	Quantity	Value	Quantity	Value	Quantity	Value
Chilli pickles	562.57	176.16	311.84	93.22	829.25	371.02	689.36	291.55
Green pickles	44.62	10.33	94.05	31.46	385.46	236.45	628.98	279.38
Mango pickles	2,685.83	835.43	2,126.88	687.24	2,227.07	733.76	2,125.49	806.69
Mango *chutney*	5,249.45	1,488.23	4,720.37	1,326.25	3,743.24	1,136.41	4,525.72	1,485.25
Tomato *chutney* & paste	685.15	180.79	443.62	129.96	1,153.51	286.84	1,999.16	566.82
Lemon *chutney*	66.41	10.80	145.73	35.32	71.10	18.46	15.88	6.20
Tamrind conc. Chutney	384.29	145.84	341.47	131.20	408.45	154.61	352.19	166.87
Other pickles, *chutneys*, pastes	3,946,51	1,274.02	5,260.57	1,984.73	5,408.26	1,959.36	5,266.02	2097.58
Soya sauce	5.65	2.16	12.20	2.05	3.48	1.50	48.57	20.79
Tomato ketchup and sauce	1,029.07	257.11	399.05	129.60	484.90	170.65	1065.91	256.15
Curry paste	102.37	46.28	208.10	112.62	183.48	101.92	82.79	59.69
Chilly sauce	2.68	1.02	3.82	1.95	88.36	48.84	24.23	11.95
Others	143.10	64.29	245.44	67.09	70.23	45.50	122.13	53.45

Source : DGCIS.

Problems

In the export of these vegetables to the international market, the exporters face several problems, they are :

(1) The entire procedure of the goods leaving the exporters, warehouse and to place on the flight, takes long hours and the first outer check point for pre-clearance, the initial and subsequent weighments and security-check procedure cause undue delay. Hence, these factors gravely affect the freshness of the produce and a certain percentage of shipped goods has to be written off by the time they reach their destination.

(2) Due to the discriminatory freight policy, the cost of transportation and the price of the products are moving upward.

(3) The available cargo space is insufficient.

(4) The storage facilities are inadequate and primitive. The quality of cargo handling at the airports is poor. There are losses due to pilferage.

So, as to promote the volume of vegetable exports and by realizing the potential, the government has recently taken up several measures. These are :

(i) continuation of Air Freight Subsidy Scheme on fresh vegetables;

(ii) establishment of integrated cargo handling facility for perishable products at Indira Gandhi International Airport, New Delhi and

(iii) opening up of a Regional Office of APEDA at Guwahati for developing and promoting exports of horticultural products. For the promotion of our vegetable exports, there is a need to formulate a suitable and pragmatic export strategy on a long-term basis; along with these, there is also the need to strenghthen the infrastructural facilities in our country.

India should try to boost export of some of the local and indigenous vegetables, which are known for their therapeutic importance and grown in abundance. There is also a tremendous scope for exporting frozen vegetables particularly to the European market. Hence, efforts are needed to solve the prevailing problems with the help of a planned strategy.

10
Vegetable Crops : Present Problems and the Future

India, being the second largest producer of vegetables in the world and has made remarkable achievement in the field of production and productivity of these, still it is insufficient to meet the ever-growing domestic and external demand. In the coming years, a major shift in consumption pattern in favour of fresh and processed vegetables is expected, hence, there is an urgent need to solve the prevailing problem of this sector and to develop this in the future. In this regard, there is the need to identify the prevailing problems of this sector and to find out the required measures to overcome these. Hence, this chapter concentrates on the problems and the future of this sector in our country. The problems of this sector can be classified into two, viz., pre-harvest and post-harvest.

I. Pre-harvest Problems

(1) Vegetables are highly susceptible to a number of diseases and severity of some of the very serious disease like feaf curl, early blight and late blight in tomato, phomopsis in brinjal, dieback in chillies, black rot in cauliflower, powdery mildew in pea etc. as result of these, the productivity and production fluctuates.

(2) *Problem of insect pests.* In tomato fruit borer, in brijal shoot and fruit borer in okra, Jassids, in chillies thrips and in cabbage diamond back moth and in cucurbits fruitfly reduce the quality of vegetables.

(3) The technologies generated in vegetable crops have hardly disseminated to the field and farmers are still unaware about the major changes which took place in various disciplines of vegetable production.

(4) *Non-availability of disease resistant hybrid variety seeds to the farmers.* Generally, the seeds of hybrid varieties are expensive and sometimes it is not easily accessible to the farmers. Due to this, the total area covered by hybrid varieties in the country is limited. Although the breeder seed production responsibility has been given to State Agriculture Universities and ICAR institutes, the seeds have not been produced as per requirement of the different organizations. Thereby, a shortage of quality seeds of improved varieties exists. As a result of all these, the productivity in India is much lower in the world.

(5) Lack of proper planning and the execution of research programmes has resulted in the disparity among states regarding productivity of vegetables.

(6) Non-availability of proper technology for growing vegetables under protected cultivation and the failure in developing a technology for growing vegetables under extreme environmental conditions like cold desert, antarctica etc.

(7) Absence of proper trainig to the farmers as well as demonstration of modern technology has resulted in poor performance in production.

(8) Research work on some of the vegetables has been totally neglected which restricted the scope for an improvement in it.

(9) With the liberalization of policy of import of new varieties/hybrids, there are chances of outbreak of some new diseases and pests.

(10) Information of plant protection is inadequate.

(11) Water management devices are inadequate and the power supply is unreliable.

(12) Difficulty in getting labourers and the cost of labour is very high.

(13) Non-availability of credit facilities especially to the small and marginal farmers.

(14) Absence of an integrated approach to the use of manure and fertilizers.

(15) Lack of proper technology and equipment for harvesting.

II. Post-harvest Problems

(1) Non-availability of technologies for handling the vegetables. Low cost technologies for post-harvest handling have not been adequately developed. Carelessness and non-application of the existing technology made the way for huge amount of loss.

(2) Inadequate knowledge of curing is a major problem. Improper curing leads to huge loss especially in vegetables like potato and onion.

(3) Lack of innovative packaging technology is responsible for huge amount of losses. An overview of status of packaging in our country shows that about 30 per cent of the marketable vegetables perish. the available methods of packing leads to spoilage in transit. It is very difficult to protect the produce against mechanical hazards involved in long road journey through wooden cartons.

(4) Absence of grading and sort in either in the field level or at the market place. The reasons for this are : (a) lack of proper standards for different vegetables, (b) lack of grading facilities, (c) fear about the disposal of low grade produce, and (d) non-availability of mechanical graders.

(5) Non-availability of proper warehouse facilities for vegetable storage is another problem of this sector. Now, vegetables are stored under open conditions or in trenches and pits resulting in huge amount of losses due to rotting and drying.

(6) The available cold storage facilities are insufficient. Moreover, most of them are under the control of privators, who actually charge higher rent and the available space in these are inadequate.

(7) Prevalence of pre-harvest contractors which leads to lower earnings.

(8) Movement of the produce from the place of origin to the place of consumption is another problem of the vegetable growers in India. Vegetables are mostly transported by road on bullock carts, tractors, trucks and also by rail and air. As a major portion of the vegetables originates from villages, these have to be transported to the markets without any delay only by road. In most of the growing areas, the roads are not proper and in certain backward areas, there is no road facility along with these absence of all-weather roads, non-availability of quickest means of transport; all of these lead to delay in transportation and ultimately result in spoilage of vegetables. Apart from these, improper handling during transportation results in huge amount of loss to the produce. The cost of transportation has been increasing in recent years. The cargo service for the movement of vegetables on the national airlines is infrequent and inadequate. There are no specific arrangements at the ports of dispatch dealing with consignments of perishable vegetables. Non-availability of cargo space leads to huge amount of economic loss.

(9) *Absence of proper weighing systems.* Most of the vegetables are sold by other than weight basis either by baskets or thorugh number; because of this type of sale, neither the buyer nor the seller will know the excact quantity and exact price.

(10) Absence of organised marketing system for vegetables is yet another major problem faced by the growers in India. Marketing of fresh vegetables faces a number of constraints due to their bulky nature, seasonality and high degree of perishability. Because of these, it leads to the involvement of a number of intermediaries, middlemen or commission agents who dominate the trade and realise huge profits. It is generally believed that the growers do not get remunerative prices for their produce, while the consumers have to pay higher prices for the same. The various costs incurred in the marketing of vegetables and the margins of profits intercepted by

different categories of market functionaries have worsened the situation. Because of all these, the prodcers are deprived of their legitimate shares in the prices paid by the consumers.

(11) Most of the growers of vegetables sell their produce in the weekly village markets either to the consumers or to the commission agents. In these weekly village markets or shandies, they are forced to sell their produce on the road side without any shelter and it leads to large volume of loss. Apart from this, the fees collected in these markets are high which reduces the scope for obtaining a major share of the consumer's price.

(12) At present, the marketing of vegetables is not regulated under the Agricultural Produce Marketing (Regulations) Act, in force in various states in India. Among the vegetables, only onion and potato, the semi-perishables are covered under the market legislation. The market fees, weighing practices, labour charges for the green and leafy vegetables are not regulated and there is no grading in the real sense in operation. Most of the wholesale markets built decades ago are congested and lack physical facilities for display, storing and weighing.

(13) Non-functioning of cooperative system of marketing except at a few places.

(14) Non-availability of adequate facilities for processing of vegetables, particularly the process of tomato, onion, garlic, etc. As a result, the crops during the peak season are not being utilized properly by the processors and thus rates become very less which has detracted the cultivation vegetables in many areas.

(15) Most of the processing industries in the country lack the required infrastructural facilities and thereby their capacity utilisation is very low.

(16) Poor export performance over the years is a threat to the sector in Inda. There are no organised efforts to produce vegetables exclusively for exports.

(17) Absence of proper planning and sound R & D support for export-oriented production.

(18) In its International marketing, these products face several problems, like wastage due to spoilage which occurs at the cargo handling stage, increasing level of packaging cost, problem of despatching from the airports, time taken for clearance reduces the freshness, higher level of transport cost, discriminatory freight policy, insufficient cargo space and inadequte storage facilities at the airport centres etc.

(19) Prevalence of stiff competition in the International market from other vegetable producing countries. This is there because our system of grading and marketing does not meet the requirement of the International market.

(20) Lack of supportive policies for vegetables by the Government reduced the scope for an increase in production and exports.

(21) Non-availability of reliable statistics on vegetable crops, on area, production, productivity and seed production, utilisation and exports.

Future of the Sector

The vegetable sector has established its credibility for improving the productivity of land, generating employment, improving the economic conditions of the farmers and entrepreneurs enhancing exports and above all providing nutritional security to the people. Even then, there appears to be the prevalence of seveal problems and challenges in this which calls for proper solutions. As the country is blessed with agro-climatic conditions suitable for year round producion of all types of vegetable, it calls for appropriate research and development efforts and policy support for its development and growth. There will be a definite shift in dietary requirements of the people with rise in income in the future, so the demand for different kinds of vegetables will go up. Assuming a populatio growth rate of 1.7 per cent and country's commintment for export, it is estimated that 2020-21 the country needs to produce about 220 million tonnes of vegetables. This has to be met through vertical growth in environment friendly manner. The increasing demand possesses challenges as well as provides opportunity for the growth of vegetables which ultimately calls for strategic planning. This plan should include the following aspects so as to obtain the ultimate goal of development in this sector. They are :

(1) So, as to control the diseases and to overcome this problem there is the need to provide proper training and guidance to the farmers on aspects like the application of fungicide and certain bio-control agents.

(2) Efforts are needed to control the vegetable pests through the use of natural enemies, parasites predators, host specific insect viruses and other enlomopathogenic micro organisms. As a whole, there is the need for an Integrated management of pest and diseases. In this regard, the existing ignorance of the growers has to be removed by popularising this by intensive efforts on research along with extension.

(3) There is the need to popularise the technologies generated in vegetable crops through education and training to the farmers. This will create an awareness among the farmers.

(4) Production of quality seed and planting material both in the public and private sector and supplying it at a lower price through authorised agencies like cooperative societies or fair price shops is urgnetly required.

(5) Proper planning and execution of research programme at regional basis is required to reduce the disparity among the states with regard to productivity since the productivity of different vegetables at farmer's field is comparatively lower than the coordinating research trials conducted in different agro-climatic conditions of the country.

(6) Efforts are needed to generate viable technology for growing vegetables under abiotic stress conditions like raising of crop under high temperature, low temperature, salinity and alkalinity. So as to increase the production of vegetables in India, there is the need to bring

additional area under these by using hybrid technology and improved agro-techniques. Another potential appoach is required under protected conditions. It involves protection of different production stages mainly from adverse environmental conditions such as extreme temperature, hail storm, scorching sun, heavy rains, etc. These conditions can be created locally by using different types of structure. These structures can be designed as per climate modification requirement of the area.

(7) For raising vegetable crops, there is the need to popularise some of the common plant protection structures such as green house, low tunnels, shading net houses, anti-hail nets, birds protection nets, wind breakers, floating cover etc.

(8) Vegetable production in green houses enables taking early crop and better management of plants. In arid regions, green houses supported with cooling device would make vegetable production possible during extreme summer. In tropical climate, green houses can provide opportunities of raising high value vegetable crops throughout the year. There is potential and opportunities in various agro-climatic zones in our country for commercial use of green houses in vegetable production. As a whole, the need exists in vegetable research institute and laboratories to standardize green house production technologies.

(9) The extension programme for transfer of vegetable production technology should be enriched and sensitized. In order to convince the farmers to adopt better technology, it requires proper training as well as demonstration of newer technology.

(10) A separate financial Apex institution should be started so as to cater the needs of perishable growers in our country.

(11) There is the need to design and implement post-harvest handling facilities and practices. The facility should include training harvest labour to handle products gently, harvesting at a proper stage of maturity. Timely handling

and field packing will be useful to keep up the quality of the produce.

(12) Vegetables should be precooled immediately after harvesting to remove field heat, if necessary, then it should be kept in refrigeration to minimise the loss. Washing of vegetables hould be done by mechanical sprayers. Waxing can also be adopted as suggested by the CFTRI.

(13) Better techniques are to be developed for convenient packing.

(14) Need to introduce mechanical graders for grading the vegetables. Proper follow-up is needed to adopt the prescribed standards. The prevailing constraints in grading have to be removed and efforts should be made to create the facilities and make it compulsory to grade the produce before it is sold.

(15) Need to provide proper storage facilities at the farm level through cooperative basis. Facilities and incentives should be given to the farmers as well as to the private storage units for improving the storage capacities. Along with these, more warehouses should be established with cold storage facilities to preserve the material during glut seasons. The present production and marketing of vegetables in our country now calls for cool chains. A cool chain is regarded as a series of storage facilities, which provide ideal conditions for preserving vegetables from the point of production to the point of consumption. This consists of pre-cooling units, refrigerated transport, cold-storage and refrigerated retail shops.

(16) The government should take steps to provide quick, efficient and cost effective transport systems. Along with these, there is an urgent need to improve the road conditions, creation of road facilities etc.

(17) Arrangements have to be made to sell the produce on the basis of weight only for all vegetables.

(18) For the elimination of pre-harvest contractors, steps like advancing production and marketing credit against

hypothecating the future crop, entering into direct contract with the processing units etc. should be taken up.

(19) There is the need to control the activities of the commission agents or middlemen for encouraging self-marketing. mere regulation of marekting by a simple legislation does not solve the problem and what is needed and lacking is strict supervision. For this purpose, fixing up of timings for auctioning different produce, grading the produce, selling by weight only etc., for orderly transactions in the market yards is an urgent requirment. In order to control the costs and raise quality, large retailers should buy the produce directly from the farm. This will enable them to greatly increase the efficiency by reducing the number of intermediaries between the farm and the retailer outlet.

The co-operatives should be developed as a real alternative channel of trade by formulating policies which will attract the growers to its fold. For this, steps like procuring the produce without prior indent, accepting all the grades of the produce, locating purchase centres at the main wholesale market places, wide publicity etc. will be helpful. The establishment of more horticultural producers cooperative marketing societies at village, Taluk and district levels and by creating marketing boards, it is possible to improve the marketing efficiency. The prevailing cooperative marketing network for horticultural products at the national level and in different states providing good servcies are the eye openers in this regard.

(20) Arrangement should be made to disseminate market information on a daily basis as regards the prices prevailing at various vegetable wholesale markets.

(21) The possibilities of fixation of floor prices for at least important vegetables should be studied and implemented by the State governments to ensure that the producers are assured of reasonable returns for their investments. For this, the existing systems need to be streamlined and the

cooperative societies' facilities are to be extended at the grassroot level.

(22) Steps should be taken to improve the shandies or weekly markets by creating basic infrastructural facilities.

(23) A new orientation needs to be given to the vegetatble processing industry to upgrade nutrition, minimise post-harvest losses, ensure remunerative returns to the growers, increase employement avenues in the countryside and generate foreign exchange earnings. In this regard, efforts are needed to increase the productivity of raw materials, contract growing, utilisation of by-products, product diversification, public policies and organised promotion. For all these, an integrated action plan is needed. Efforts are also required to popularise the processed products both in the domestic and external markets through demonstrations, advertisements etc. So, as to improve the capacity utilisation and the required basic facilities in this regard.

(24) Proper planning for export oriented production is necessary for the quality of produce with standard packing at a competitive price during the required period. For this, a sound R&D support is essential.

(25) Constraints like inadequate cargo space, high air freight, poor handling and storage facilities at harbours and airports have to removed.

(26) Chemical residue problem in fresh vegetables has to be got over to stabilise the export trade.

(27) A sustaining export trade has to be built on a reliable production system. Production exclusively for export has to be organised with safeguard against adverse climatic conditions.

(28) There is the need to generate sufficient research and developement support for popularising organic farming in scientific way which can push up the volume for our exports.

(29) Need to identify new markets in the International level for our vegetables and processed products.

(30) There is the need to collect reliable statistics on vegetable crops, on area, production, productivity, seed production which will be useful in the preparation and implementation of plans and also in the utilisation of vegetables.

(31) Need to establish a Vegetable Technology Mission by the Central Government so as to provide a nodal agency to take care of pre- and post-harvest aspects. This will increase the production as well as consumption both in the rural and urban areas.

(32) To meet the challenges of liberalisation and WTO regimes, there is the need for a national seed policy so as to promote healthy development of both public and private seed sectors.

(33) There is the need to provide proper incentives and facilities to the corporate managements since they are assisting the growers through soil testing, providing quality seeds of high yielding varieties and other inputs, extension services and financial assistance. Apart from these, they also halp in transportation by providing packing materials and in marketing directly to the consumers throgh co-operative societies. They open procurement centres in major production areas for collecting the raw material and help in the setting up of small scale intermediate processing units.

(34) The governement must evolve a specific crop insurance scheme so as to provide relief to growers since they invest their small savings in annual vegetable crops, and the scope for loosing their entire savings is more in this because of natural calamities which destroy the crops.

As vegetables are rich sources of carbohydrate, minerals, vitamins, proteins besides having medicinal value and provide nutritional security to a predominately vegetarian country and these crops are more remunerative and generate more employment, along with these, the demand for vegetables is increasing day by day owing to increase in population; so what is needed is, a long term strategy for a further increase in the vegetables production in our country.

11
Organisations in the Development of Vegetable Crops

Ever since the agricultural departments came into existence in India, the emphasis has continued to be on crop plants and the research on vegetables did not receive sufficient attention. During the forties and the fifties, however, with the sanctioning of a nucleus plant-introduction scheme by the Indain Council of Agricultural Research (ICAR), the evaluation of collections of indigenous material and foreign intorduction resulted in the selection of good varieties in several vegetables.

Research on the improvement of vegetable crops was initiated in 1947-48 at the Indian Agricultural Research Institute, New Delhi, and in different states like Punjab, Uttar Pradesh, West Bengal, Maharashtra, Himachal Pradesh, Jammu and Tamil Nadu under adhoc schemes sponsored by the ICAR. the Government of Inida established the Vegetable Breeding Station at Katrain in Kulu Valley (Himachal Pradesh) in 1949, which was transferred to the Indian Agricultural Research Institute in 1955 which has been carrying out intensive research on temperate vegetable and their seed production. After the establishment of agricultural Universities in different states, work on the improvement of vegetable crops was taken up in these universities and also at the Indian Institute of Horticultural Reserarch (Karnataka), which was established in 1968 and the IARI set up a separate division of vegetable crops and floriculture in 1970. In 1982, the latter was further divided. Apart form these IARI, IIHR and their regional stations vegetable improvement programme is also being carried out at the Central Institute of Horticulture of Northern plains, Lucknow; Vivekanada Parvatiya Krishi Anusandhanshala, Almora, Central Agricultural Research complex for the NEH region, Shillong. During the Fourth Five Year Plan period, an All India Co-ordinated vegetable improvement project (AICRP) was started by the ICAR during 1970-71 to provide a national grid for testing the technologies developed by various research institutes and agricultural universities through inter-disciplinary multilocational research approach. the project was started with 7 main and 10 sub-centres; three more centres were added to it during the Fourth, Two in the Sixth and Two in the Seventh Plan. After the upgradation of this project in the Seventth Plan, at present, it is undertaken as multi-disciplinary, multi-location research at 23 centres, along with 27 voluntary centres (Table 11.1).

At present, research and developement on horticultural crops is being undertaken in the country in 10 ICAR institutes, 1 project directorate, 10 national research centres, 16 All India coordinated research projects with 215 research stations, 1 full fledged university of horticulture and 25 state agricultural universities.

The objectives of research are :

(1) Incorporating resistance against major diseases and insect-pests.

(2) Developing superior cultivars capable of giving high quality-yield.

(3) Heterosis breding.

(4) Standardisation of agrotechniques for increasing productivity.

(5) Disease and insect-pest management.

(6) Post-harvest management with a view of reducing the post-harvest losses in selected vegetable crops.

In the national research programme on vegetable crops, the ICAR at various central institutes and state agricultural universities and others included 25 vegetable crops, namely brinjal, tomato, chilli, sweet pepper, cabbage, cauliflower, cowpea, carrot, french bean, okra, peas, onion, watermelon, muskmelon, amaranthus, bittergourd, pumpkin, radish, squash, pointed gourd, roundgourd and garlic.

Table 11.1 : Research Centre of AICRP on Vegetable Crops

Institution	Location	State	Year established	Mandate crops
Project Directorate of Vegetable Research	Varanasi	Uttar Pradesh	1993	Brinjal, cauliflower, chillies, cowpea, cucumber, peas, muskmelon, tomato, okra, cabbage.
Indian Institute of Horticultural Research	Hessarghatta	Karnataka	1970	All vegetable crops
Indian Agricultural Research Institute	New Delhi	Delhi	1970	All vegetable crops
Indian Agricultural Research Institute Regional Station	Katrain	Himachal Pradesh	1970	All vegetable crops
Andhra Pradesh Agricultural University	Hyderabad	Andhra Pradesh	1988	Tomato, cucurbits, melons, french beans
Andhra Pradesh Agricultural University	Lam	Andhra Pradesh	1970	chillies
Assam Agricultural University	Jorhat	Assam	1970	Tomato, cabbage, melons, french beans
Rajendra Agricultural University	Sabour	Bihar	1970	All vegetable crops
Gujarat Agricultural University	Junagarh	Gujarat	1988	Tomato, okra, onion, garlic
Haryana Agricultural University	Hissar	Haryana	1975	Tomato, okra, cucurbits, cauliflower
Dr. Y.S. Parmar University for Horticulture and Forestry	Solan	Himachal Pradesh	1970	Tomato, french bean, capsicum, cauliflower.
Sher-e-Kashmir University for Agriculture and Technology	Srinagar	J&K	1970	Cole crops, tomato, capsicum
University of Agricultural Sciences	Dharwar	Karnataka	1994	Brinjal, bitter gourd, bottle gourd, chillies, cucumber, garlic, tomato, okra.
Kerala Agricultural University	Vellanikkara	Kerala	1989	Tomato, brinjal, chillies
Jawaharlal Nehru Krishi Vishwa Vidyalaya	Jabalpur	Madhya Pradesh	1970	Toomato, garden, pea
Indira Gandhi Krishi Vishwa Vidyalaya	Raipur	Madhya Pradesh	1994	Brinjal, chillies, garlic, onion, cabbage, tomato, musk-melon, peas.

contd...

Table 11.1 – Contd...

Institution	Location	State	Year established	Mandate crops
Mahatma Phule Krishi Vidyapeeth	Rahuri	Maharashtra	1970	All vegetable crops
Marathwada Agricultural University	Ambajogai	Maharashtra	1975	Tomato, onion, water melon.
Orissa University of Agriculture and Technology	Bhubaneshwar	Orissa	1970	Tomato, okra
Punjab Agricultural University	Ludhiana	Punjab	1970	All vegetable crops
Rajasthan Agricultural University	Durgapura	Rajasthan	1970	Tomato, melons, pea, cauliflower
Tamil Nadu Agricultural University	Coimbatore	Tamil Nadu	1970	All vegetable crops.
Narendra Dev University for Agriculture Technology	Faizabad	Uttar Pradesh	1988	Tomato, cucurbits, melons, and french beans
Chandra Shekhar Azad University of Agriculture and Technology	Kanpur	Uttar Pradesh	1970	Tomato, pea, cauliflower, onion chillies.
Gobind Ballabh Pant University of Agriculture and Technology	Pantnagar	Uttar Pradesh	1970	Tomato, brinjal, french bean
Bidhan Chandra Krishi Vishwa Vidyalaya	Kalyani	West Bengal	1975	Tomato, Okra, cucurbits, cauliflower

Research Achievements

(1) Many exotic crops such a s gherkin, baby corn, broccoli, brussel's sprouts, asparagus, celery etc. were introduced for commercial cultivation.

(2) 181 improved cultivars were released including 40 F1 hybrids.

(3) Tropical seedless watermelon is released.

(4) Off-season cultivation of tomato, cauliflower and onion was standarised.

(5) Use of chemicals for increasing productivity was standarised.

(6) Agro-techniques for protected cultivation of vegetables were evolved.

(7) Developed periodic models.

(8) IPM in cabbage and cauliflower and tomato fruit borer were developed by using mustard and marigold as trap crops respectively.

(9) Eco-friendly, low cost storage system at farm gate for vegetables, potato and onion was developed.

(10) Development of packages for vegetables.

Future Needs

Future reseach strategies on vegetables should focus attention on the following aspects. They are :

(1) Need to undertake innovative basic, strategic and applied research for developing technology to enhance productivity of vegetable crops.

(2) Need to disseminate the production technology to the farmers and to provide consultancy in vegetable research and developement.

(3) Need to develop hybrids resistant to multiple diseases and pests.

(4) Need to identify varieties suitable to protected conditions.

(5) Need to undertake studies on production technologies of vegetables.

(6) Need to develop technology on organic farming.

(7) A detailed study is needed to assess the post harvest losses in vegetable crops and a suitable planned strategy for reducing the losses has to be formulated without much delay.

(8) As a whole, there is the need ot undertake state-wise and item-wise survey on various aspects of vegetables, viz., pre and post-harvest.

Central Potato Research Institute (Shimla)

It was started in 1934 and became a full-fledged institute in 1949. Since then, it is putting its efforts to improve the well being of potato growers and its consumers through its various research programmes. At present, it has 7 regional stations at Kufri (Himachal Pradesh), Jalandhar (Punjab), Patna (Bihar) Shilong (Meghalaya), Madipuram (Uttar Pradesh), Gwalior (Madhya Pradesh) and Ootacamand (Tamil Naud).

Objectives

(1) To undertake basic and strategic research for developing technolgies to enhance productivity and utilization of potato.

(2) To produce disease free basic seeds of different notified varieties developed by the institute.

(3) To act as national repository of scientific information relevant to potato.

(4) To provide leadership and coordinate network research with state agricultural universities for generating location and variety specific technologies and for solving area-specific problems of potato production.

(5) To collaborate with natioal and international agenecies in achieving the objective.

(6) To act as a centre for training in research methodologies and technology for upgrading scientific manpower in modern technologies for potato production.

(7) To provide consultancy in potato research and development.

Since the inception of the institute, the emphasis in potato research has been on the development of high yielding varieities and in this regard, it has released 34 high yielding varieties. Along with these, standardization of packages of practices for such varieties and increasing the availabnility of disease free seed stocks for sustaining the explosive growth in production was developed which brought in revolution in the potato. However, certain areas have received less importance, which needs to be tackled. They are :

(1) Strengthening of existing germplasm resources and their better utilisation.

(2) Need to develop varieties and agro-techniques that would enable the spread of potato to non-traditional areas and other seasons.

(3) Genotypes with high tuber dry matter and low reducing sugar accumulation during storage need to be developed.

(4) Need to develop varieties with good keeping quality so as to reduce the dependence on energy intensive refrigerated storage.

(5) Need to develop low-input, sustainable and environment-friendly agro-techniques which reduce the dependence on chemical fertilizers, toxic chemicals and systemic insecticides etc.

Central Tuber Crops Research Institute (CTCRI)

It was started in 1963 and has regional centre at Bhubaneswar and an All India Coordinated Project on Tuber Crops. The CTCRI is the only one of its kind in the world carrying out research and development exclusively on tuber crops. The objective of the institute is to undertake basic, strategic and applied research on production and processing aspects of tuber crops.

At present, a rich stock of genetic resources of tuber crops are there at CTCRI which maintains more than 3500 accessions;

1500 of cassava; 800 of sweet potato; 600 of yams; 450 of aroids and 150 of other tuber crops.

Research Gaps

The future research strategy should concentrate on evolving high yielding, short duration, starchy and high quality varieties in tuber crops, development of low input produciton technolgoy, management of cropping system involving tuber crops, biosafety aspects, diversification for alternative uses, farmers participatory research for generation of user friendly technologies, agrotechniques for non-traditional areas, control of pollution problem of tuber crops base industries, processing equipment and structures for efficiency enhancement, market assessment and price behaviour of tuber crops etc. All of these call for a perspective plan and with the help of this, it is possible to improve the production and utilisation of these crops, besides uplifting the economic status of the growers.

National Research Centre for Onion and Garlic (NRCOG)

The ICAR has started NRCOG with mission made approach at Nasik during the Eighth Plan. The centre was established in 1994 with its headquarters at Nasik and in 1998 it has been shifted to Rajgurunagar, Pune. The institute was started with the following objectives :

(1) To collect, maintain and act as national repository for onion and garlic germplasm.

(2) To develop hybrids suitable for domestic as well as export market coupled with resistance to biotic and abiotic stresses.

(3) To enhance and sustain productivity and quantity of seeds as well as bulb crops through agronomic manipulations.

(4) To develop package for post-harvest handling and value addition.

(5) To act as clearing house of research and general information related to onion and garlic.

Future Requirements

The future research plan of NRCOG should have the following features :

(1) Increasing the productivity of improved varieties through better agronomic manipulations.

(2) Loss prevention through genetic manipulations, cultural management and improvment in storage and packaging conditions.

(3) Value addtion through processing of the bulbs in the form of dehydrated flakes, dehydrated powder, paste etc.

(4) Development of export-oriented varieties.

(5) Finding alternative technologies for an extension of onion cultivation in non-traditional areas.

(6) Promotion of export of onion seed.

Defence Reseach and Development Organisation (DRDO)

It has established low cost green houses in the cold desert areas of Ladakh where several individual farms have set-up small units with improved design for growing vegetables during the off-season. This concept is now spreading to the sub mountain areas of Himachal Pradesh, Uttar Pradesh and Jammu and Kashmir.

Apart from these, most of the export-oriented units are engaged in green house cultivation which are set up in collaboration with multinationals and have started production.

Marketing Research Institutions

The importance of an efficient marketing system was first recognised by the Royal Commission on Agriculture in its report in the year 1928. It had recommended for the appointment of marketing experts in the Agricultural Department of all the major provinces. Till that time, there was no official agency directly concerned with this basic activity while endorsing the recommendations. The Central Banking Enquiry Committee (CBEC) in 1931, recommended for the setting up of a Central agency for initialling and co-ordinating state activities relating to the development of agricultural marketing. From then onwards, several marketing institutions are engaged in conducting agricultural marketing research in India. A brief note on these is given below in this section.

(1) Directorate of Marketing and Inspections

On the basis of the recommendations of the CBEC, the Government of India set up a Central Organisation in 1935 known as "The Office of the Agricultural Marketing Adviser to the Government of India". The organisation later expanded and was renamed as Directorate of Marketing and Inspection. Since its inception, it continues to be responsible for bringing about an integrated development of marketing of agricultural produce with a view to safegaurd the economic interest of the producers-sellers as well as that of the consumers.

Functions

  (1) Statutory regulation and development of markets.
  (2) Promotion of grading and standardization of agricultural and allied commodities.
  (3) Market Research, Survey and Planning.
  (4) Market Extension.
  (5) Training of Personnel.
  (6) Administration of Cold Storage Order, and Food Products Order.

(2) State Agricultural Marketing Boards

These are there for the purpose of supervising and advising the functioning of the regulated markets. It is an exclusive-cum-advisory body. It affects improvements in the regulation scheme supervises the functioning of regulated markets and adivses market committees and the State Government on related matters.

(3) Indian Council of Agricultural Research (ICAR)

It encourages research in agricultural universities and it own institutions on agricultural marketing by financially assisting them in their research projects.

(4) *National Council of Applied Economic Research (NCAER)*

It conducts studies on agricultural makreting which are of all India significance.

(5) *Agricultural and other Universitites*

They too are engaged in conducting marketing studies through their departments of Economics, Agricultural Economics, Agricultural Marketing and Horticulture.

National Horticulture Board (NHB)

The Government of Inda, Minstry of Agriculture set up the NHB in 1984 for promoting integrated development of Horticulture in the country. It envelopes all the aspects of horticulture development in the country, the thrust area being the pot-harvest management and marketing. It has various schemes for the exploitation of comercial horticulture, post-harvest management and marketing to provide support services like market information and transfer of technology and also schemes to induce awareness about innovative ideas and concepts.

The NHB is collecting market-related information on sales prices and market arrivals of vegetables and fruits from 33 centres located all over inida called NHBNET. These centres collect the market information and send it to a coordinating cell for publication of monthly bullentins. It has aided surveys for various states and Union territories for the development of horticulture. In order to address the problem of malnutrition, it took up the programme of establishement of nutritional gardens in rural areas.

It provides finanacial assistance to the State Governments for conducting visits to various research stations to familiarize the horticultural growers with latest agro-techniques. Its network is proposed to be increased ot cover more centres in the Ninth Plan.

The NHB under the "Integrated Project for Management of Post-harvest Infrastucture of Horticultural Crops" has established 127 grading/packing sheds, 64 pre-cooling units/short duration cool stores, 39 cold storages, 76 specialised transport vehicles, 53 refrigerated transport vehicles, 214 retail outlets and 11 auction platforms and has provided subsidy for plastic crates.

National Co-operative Development and Warehousing Board (NCDC)

This was started in 1956 and in 1963; the Board was converted into the National Co-operative Development Corporation. It is engaged in the marketing, processing or storage of agricultural produce and advances loans and grants to State Governement for financing co-operatives engaged in the above stated activities. This is engaged in creating infrastructure for post-harvest management of vegetables too. It has sanctioned several projects to improve the marketing and reduce post-harvest losses. Almost the entire cold storage capacity in the cooperative sector has been created with the technical and financial assistance of NCDC in the country up to March 1998, 250 cold storages with a capacity of 7.71 lakh tonnes have been assisted of which 239 cold storages with a capacity of 6.97 lakh tonnes have been installed. The corporation approved 167 projects in Maharashtra for the establishment of pre-cooling and cold storage units. It has also assisted a marketing project to be implemented by HIMFED, Shimla, and Lahaul Potato Grower's Cooperative Society, Manali.

Central Warehousing Corporation (CWC)

It was established as a statutory body in 1957. Its main functions include building of godowns and warehouses, running warehouses for storage of agricultural produce, seeds, fertilizers etc. It is also running air-conditioned godowns at Kolkata, Mumbai, Delhi and provided cold storage facilities at Hyderabad.

The National Agricultural Co-operative Marketing Federation (NAFED)

It was established in October 1958. The State Level Marketing Federations and the National Co-operative Development Corporation are its memebers. the head office of it is at New Delhi, and branch offices are located at Mumbai, Kolkata, and Chennai. Its area of operation extends to the whole country. The major objects of NAFED are :

(a) To co-ordinate and promote the marketing and trading activities of its affilitated Co-operative Institutions;

(b) To make arrangements for the supply of the agricultural inputs required by member institutions;

(c) To promote Inter-State and International trade in agricultural and other commodities; and

(d) To acts as an agent of governement for the purchase, sale storage and distribution of agricultural products and inputs.

Activities

(1) NAFED is engaged in Inter-State trade in agricultural commodities with a view to getting better prices for the producers.

(2) NAFED is exports agricultural commodities, which includes onion, potatoes, processed vegetables, ginger, chillies, garlic etc., to various countries.

(3) NAFED arranges for the import of agricultural products and inputs.

(4) NAFED maintains expert staff which conducts market studies, collects useful data and circulates the results among the members.

Apart from these, it maintains contacts with the input supply institutions, such as National Seed Corporation and the Fertilizer Corporation of India.

Other Agencies

Besides NAFED, about 12 State and Central level societies and 275 Primary Marketing Societies are directly engaged in the marketing of vegetables. Other cooperative societies providing good services in the marketing of vegetables are Horticultural Produce Marketing, Nilgiris Cooperative Marketing Society, Nilgiris Vegetable Growers Coperative Marketing Society, Udhagamndalam etc. In Karnataka, the cooperative societies have expanded on a large scale with the establishement of district level cooperative societies. There are about 54 registered cooperative societies in the State. As a whole, in its marketing, NCDC is also playing an important role in our country. The other notable cooperatives are

Fresh in Hyderabad, Apni Mandi, Several State Trial Cooperative Corporations headed by TRIFED etc.,

Agricultural and Processed Food Products Export Development Authority (APEDA)

It was established in 1986 with the aim to boost Indian agro-based products through optimum utilization of the enormous resource base of the country.

Objectives

(1) Development and subsistence of global competition to the Indian agro-products.

(2) To provide better income to the farmers through higher unit value realisation.

(3) To create employment opportunities in rural areas by increasing value added exports of farm products.

For achieving these objectives, it has identified new markets, provides better support systems to the exporters and manufacturers and introduces new products to the international market. Fresh vegetables and processes vegetables are coming under its purview.

APEDA undertakes several developmental programmes to enhacne the exports which are :

(a) Developement of database on products, markets and services.

(b) Publicity and information dissemination.

(c) Invites officials and business delegations from abroad.

(d) Organisation of product promotions abroad and visits of officials and trade delegations aborad.

(e) Participation in International Trade Fairs in India and abroad.

(f) Organisation of buyer-seller meets and other business interactions.

(g) Information dissemination through APEDA's newsletter, feed back series and library.

(h) Distribution of Annual APEDA Awards.

(i) Provides recommendatory advisory and other support services to the trade and industry.

(j) Problem solving in government agencies and organisations.

Apart from these, under the garb of its various schemes, it offers the required financial assistance to tap the vast export potential in the agricultural sector. It also undertakes market surveys for penetration of product range in new markets. It has schemes for export promotion and market development, for packaging development, for promoting quality and quality control, for organisation building and human resource development, for generating relevant research and development by APEDA through research institutions, for infrastructure development etc.

CFTRI

This is research institute under the CSIR working on horticultural crops for their preservation and processing.

Marketing of Seeds by Private Sector

The main agencies involved in this are Maharashtra Hybrid Seeds Jaina, Indo-American Hybrid Seeds, Bangalore; Proagro Seeds, New Delhi; Century Seeds, Delhi; Namdhari Seeds, Bangalore; Sungrow Seeds, New Delhi; Ankur Seeds, Nath Seeds, Aurangabad; Beejo Sheetal, Jaina; Sandoz; Suttons Seeds, Kolkata; Pioneer Seeds, Bangalore Nurserymen Cooperative society etc.

All India Coordinated Research Project in Post-Harvest Technology of Horticultural Crops

This was started in 1978 to carry out research on the post-harvest physiology of fruits and vegetabels with the objective of retaining their quality and minimising losses during handling, transport, storage and processing. At present, there are 13 Research Centres and 6 Voluntary Centres under this. The objectives of this are :

(1) Improvement of pre-harvest and harvesting techniques to maximise quality and minimise post-harvest losses.

(2) Reduction of post-harvest losses and maintenance of quality characteristics by improving systems for pre-cooling, handling, packaging and transport.

(3) Development of appropriate storage techniques to minimise losses and prolong quality characteristics.

(4) Processing for maintaining quality components during storage and marketing.

(5) Utilisation of wastes from commercially unacceptable vegetables and processing systems.

Over these years, it has achieved a lot on the above aspects and come up with certain acceptable solutions like the development packaging technology, on farm storage and technology for preparation and preservation of various products from tomato, black carrot etc.

The above aspects on the role of different organisations or institutions on pre- and post-harvest activities on vegetables clearly show that there appeared a remarkable success. However, considering the future needs, the past researches are insufficient in both the aspects. As far as pre-harvest aspects are concerned, the scope for improvement is there on the following aspects. They are :

(1) Improving the productivity.

(2) Work need to be intensified on developing varieties with wide adaptability and resistance to several important diseases and insect-pests in tomato, brinjal, okra, chilli, onion, cole crops, peas and bean.

(3) need to develop good hybrids in cauliflower, cabbage, tomato, onion etc. and supply it to the farmers.

(4) Need to breed procesing qualities in tomato, onion, carrot, peas etc.

(5) Research on off-season vegetable production.

(6) Work on under exploited vegetables.

(7) Development of vegetable-based cropping systems.

Post-harvest Gaps

Works on post-harvesting activities undertaken in our country is below the requirements. The attention given by either the organisations or the privators are insufficient. Hence, the scope for this is vast. In this regard, the following areas have to be taken into consideration.

(1) Improving the packaging technology.

(2) Developing the most economic and convenient methods of transport system.

(3) Finding the most economical efficient methods of storage. Need to carry out regulatory as well as promotional and developmental activities for developing short-term storage, designing for storage of vegetables in rural areas, to evolve commodity-wise methodology for scientific preservation, to prepare state-wise micro level plan to identify areas of demand, to collect cold stoage data and to organise seminars and workshops on cold storages.

(4) Improvement in the processing activities.

(5) Improving the marketing efficeincy.

(6) Estimation of marketing costs, margins and price spread for different vegetables.

(7) Estimation of the benefits of a market regulation programme.

(8) Projection of the demand for and supply of vegetables and gaps in different periods both in the micro and macro level.

(9) Estimation of the extent and magnitude of price fluctuations for different vegetables and the factors affecting them and the nature of price movements in different types of markets.

(10) Export oriented research on different vegetables.

(11) Data-base of reliable statistics on vegetable crops, on area, production, productivity and seed production should be established to help perspective planning of production, consumption and exports.

Publications of Production and Marketing Information

Pre- and Post-harvest informations are available in the publications brought out regularly as well as periodically by the concerned departments. Some of the publications in which these informations are regularly published are :

(i) Agricultural Situation in India.

(ii) Productivity.

(iii) Agriculture and Industry Survey.

(iv) Agricultural Marketing.

(v) Indian Horticulture.

(vi) India farming.

(vii) Fertilizer News.

(viii) Agri News.

(ix) Agro India.

(x) Facts For You.

(xi) Bulletin on Prices.

(xii) India Journal of Agricultural Marketing.

(xiii) Commodity Survey Reports.

(xiv) New Papers – Economic Times, The New Financial Express and The Hindu Business Line.

(xv) Annual Reports brought out by the concerned departments, stating their progress.

(xvi) The Hindu Survey of Indian Agriculture – Annual.

Programmes and Schemes in the Ninth Plan

In the Ninth Plan, several areas have been identified for special emphasis. They are :

(a) use of plastics for micro-irrigation, fertilization, green houses etc. Encouragement will be given for technical support for efficient functioning, farm participatory demonstration on fertigation and micro-irrigation.

(b) Integrated development of vegetables including tuber crops. This scheme provides assistance for replacement of traditional seeds with hybrid seed area expansion in non-traditional region. There is emphasis on production of off-season vegetables.

(c) Scheme for post-harvest management, marketing and exports. Under this, collection and dissemination of market related information would be taken up. Financial assistance to professional organisations will also be enhanced for the development of horticulture.

REFERENCES

Abbott, J.C., *Marketing Problems and Improvemet Programmes*, FAO Rome 1958.

Albert, F., Hill, *Economic Botany*, McGraw Hill, 1952

Bose, T.K. and M.G. Som, *Vegetable Crops in India*, Naya Prakash, Calcutta, 1986.

Brunk, M.E. and L.B. Darroh, *Marketing of Agricultural Products*, The Ronal Press, 1955.

Chadha, K.L., *Towards a Horticultural Revolution*, Agriculture Today, March-April, 1998.

Choudhury, B., *Vegetables*, National Book Trust, 1998.

Das, P.C., *Vegetable Crops of India*, Kalyani Publishers, Ludhiana 1997.

ICAR, *Handbook of Agriculture*.

ICAR, *50 Years of horticultural Research*.

Jasdanwalla, Z.Y., *Marketing Efficiency in Indian Agriculture*, Allied, Bombay, 1966.

Singh, H.P., Harnessing the Potential of Horticulture, *Fertilizer News* vol. 45 (2) 2000.

Vigneshwar, V., Dynamics of Fruits and Vegetable Marketing in India, *Indian Journal of Marketing*, vol. XVII (6) 1986.

Vigneshwara, V., Marketing of Horticultural Products, *Facts For You*, Oct., 1990.

Vigneshwara, V., Fruit Exports : Time is ripe to cut a bigger slice, *Financial Express* 24 Aug., 1991.

Vigneshwara, V., Potato : Swelling Demand, Shrinking output, *Financial Express* 28, Dec., 1991.

Vigneshwara, V., Low Productivity Planges Cole Crops, *Financial Express* 8, Feb., 1992.

Vigneshwara, V., Onions : Exports Need Boost, *Financial Express* 29, Feb., 1992.

Vigneshwara, V., Cassava : Untapped Potential, *Financial Express* 4 April, 1992.

Vigneshwara, V., Low Yield Plauges Chilli, *Financial Express* 1 August, 1992.

Vigneshwara. V., Cucurbits : Growing Commerical value, *Financial Express*, 19 Sept., 1992.

Dailies

(1)   *The Economic Times.*

(2)   *The Hindu Business Line.*

(3)   *The New Financial Express.*

Journals

(1)   *Agro India*

(2)   *Agriculture Situation in India*

(3)   *Indian Horticulture*

(4)   *Indian Journal of Agricultural Marketing.*

(5)   *Plant Horti Tech.*

(6)   *Productivity.*

Annuals

(1)   *The Hindu survey of Indian Agriculture* – 1988 to 2000.

(2)   ICAR.

Documents

(1)   ICAR – *Vision 2020* – PDVR 1997.

(2)   ICAR – *Vision 2020* – CTCRI 1997.

(3)   CPRI Shimla – *Potato Research in India : A success story of Fifty Years Nov. 1999.*

(4)   ICAR – *Vision 2020* – NRCOG 1997.

INDEX

www.ingramcontent.com/pod-product-compliance
Lightning Source LLC
Chambersburg PA
CBHW072249210326
41458CB00073B/911